走进乐器

An Introduction to
Chinese and
Western Musical Instruments

主编 / 朱文利　施　维

审稿 / 郑玉章

编者 / 朱文利　施　维　周　思　易　雪
　　　　喻　梦　杨潇鸣　刘　逸　冯　婷

重庆大学出版社

内容简介

本书是一本把英语语言学习和音乐专业知识紧密结合的、系统性地分学科和分专业介绍音乐院校的主要专业乐器的音乐英语教材。本书由中国民族乐器和西方乐器两大部分组成，其中选取的皆为音乐学院开设的器乐专业的乐器。每个专业乐器的内容均包括该乐器的发展史、结构、演奏技巧、经典曲目、专业术语和练习等部分。本书旨在指导学生通过对本专业乐器相关知识的学习，掌握本专业常用的专业术语的英语表达。本书在每个单元设置了与单元内容和思政内容密切结合的综合训练题，在巩固音乐英语知识的同时引导学生提高思辨能力、增强文化自信和树立正确的价值观。

图书在版编目(CIP)数据

走进乐器：英文 / 朱文利，施维主编. --重庆：
重庆大学出版社，2023.8
ISBN 978-7-5689-3888-4

Ⅰ.①走… Ⅱ.①朱… ②施… Ⅲ.①乐器—英语—高等学校—教材 Ⅳ.①TS953

中国国家版本馆CIP数据核字(2023)第083704号

走进乐器
An Introduction to Chinese and Western Musical Instruments

朱文利　施维　主编

责任编辑：杨琪　　版式设计：孙婷
责任校对：关德强　　责任印制：赵晟

*

重庆大学出版社出版发行
出版人：陈晓阳
社址：重庆市沙坪坝区大学城西路21号
邮编：401331
电话：(023) 88617190　88617185（中小学）
传真：(023) 88617186　88617166
网址：http://www.cqup.com.cn
邮箱：fxk@cqup.com.cn（营销中心）
全国新华书店经销
重庆升光电力印务有限公司印刷

*

开本：889mm×1194mm　1/16　印张：17　字数：446千
2023年8月第1版　2023年8月第1次印刷
ISBN 978-7-5689-3888-4　定价：69.00元

本书如有印刷、装订等质量问题，本社负责调换
版权所有，请勿擅自翻印和用本书
制作各类出版物及配套用书，违者必究

前言

在音乐院校的英语教学中，音乐英语是连接英语和音乐专业之间的一座桥，而音乐英语教材则是帮助学生顺利走过这座桥的关键。音乐英语能帮助音乐专业学生听懂英语讲座、阅读与专业相关的英文文献、进行国际学术交流等，从而提升音乐专业学生学习英语的兴趣。

复旦大学蔡基刚教授说："专门用途英语教材不是纯粹地为学习语言而编写的语言教材，而是旨在引导学生在学习学科知识的同时习得语言"。本着学科知识和语言知识相辅相成的宗旨，本书编写团队花费大量时间和精力调研了音乐专业学生对英语学习的实际需求，在调研的基础上确定了编写理念。经过三年多教学实践的反复打磨，《走进乐器》一书的编写工作终于顺利完成。

与国内现有的音乐英语教材相比，《走进乐器》有以下几个创新点：

1. 突出立德树人的教材建设宗旨。本书在章节内容的编写和课后练习的设计中都融入了思政元素，让学生在学习中，不仅能收获知识，还能提高思辨能力，树立正确的价值观，并增强文化自信。

2. 系统地介绍中西方主要乐器。《走进乐器》由中国民族乐器和西方乐器两大部分组成，其中选取的皆为音乐院校开设的器乐专业的乐器。内容涵盖乐器的结构、发展史、演奏技巧、经典曲目、专业术语等，并且每个章节后设置了配套练习，方便学生既学且练。

3. 探索中国传统乐器的英文表达。目前国内外介绍中国传统乐器的英文资料非常有限，本书编写团队结合多轮教学实践，专门对如何译介中国传统乐器，尤其是乐器的结构和演奏技法做了深入的、系统性的探索。

4. 有效解决音乐英语专业术语表达混淆不清的问题。国内现有的音乐英语教材通常只有词汇总表，而没有每个乐器单独的词汇表。然而在实际学习中，中、英文的术语互译却时常不是一一对应的，常常会遭遇"一对多"或者"多对一"的情况，这就给学习者带来极大困惑。为了帮助学习者准确掌握信息，避免产生混淆，《走进乐器》不仅有词汇总表，还在每种乐器后附有该乐器专门的术语词汇表，以方便学习者使用。

  5. 个性化选取教学内容。本书的编写体系既独立又相互关联，能满足不同专业、不同学科的学生在学习过程中对知识的个性化需求，比如，各专业院系学生可选择和本专业相关的乐器知识作为学习内容。

  《走进乐器》由四川音乐学院基础部的音乐英语教学团队集体打造。其中四川音乐学院施维主要负责中提琴、大提琴、倍低音提琴、电子琴、笛子和古筝等乐器的编写工作以及本书中所有中国民族乐器在教学实践后的修订工作；四川音乐学院周思负责手风琴、二胡和古琴的编写工作；四川音乐学院易雪负责圆号的编写以及词汇的梳理工作；武汉音乐学院英语教研室喻梦负责双簧管的编写工作；武汉音乐学院英语教研室杨潇鸣负责巴松的编写工作；四川音乐学院刘逸和冯婷负责打击乐器的编写工作，其他部分皆由四川音乐学院朱文利编写。

  《走进乐器》是音乐英语教学团队编写的"音乐英语"系列教材之一。因为声乐部分的教材建设和教学实践已经很成熟，所以为方便声乐专业学生学习，暂时把声乐部分放在本书附录一；待戏剧和舞蹈部分成熟后，再将声乐部分编在"音乐英语"系列教材之二《声乐、戏剧和舞蹈介绍》中。

  非常感谢四川音乐学院院领导、教务处领导以及基础部领导对音乐英语教学团队教学改革工作的大力支持，同时非常感谢对本书中涉及的专业知识提供过指导和帮助的各位老师和同学。

编 者
2023 年 2 月

CONTENTS

Part One   Chinese Musical Instruments

Lesson One   Wind Instruments — 2
  Di — 3
  Suona — 8
  Sheng — 12

Lesson Two   Plucked String Instruments — 18
  Pipa — 19
  Ruan — 26
  Guqin — 34
  Guzheng — 42
  Konghou — 50

Lesson Three   Bowed-String Instrument — 56
  Erhu — 57

Lesson Four   Hammered-String Instrument — 64
  Yangqin — 65

Part Two   Western Musical Instruments

Lesson Five   Keyboard Instruments — 74
  Piano — 75
  Accordion — 89
  Electric Organ — 102

Lesson Six   Stringed Instruments   *107*

     Violin   *108*

     Viola   *121*

     Cello   *126*

     Double Bass   *132*

     Classical Guitar   *137*

Lesson Seven   Brass-Wind Instruments   *146*

     Trumpet   *147*

     Horn   *154*

     Trombone   *160*

     Tuba   *167*

Lesson Eight   Woodwind Instruments   *172*

     Flute   *173*

     Clarinet   *180*

     Saxophone   *191*

     Oboe   *200*

     Bassoon   *206*

Lesson Nine   Percussion Instruments   *214*

Lesson Ten   Synthesizer   *226*

Appendix I   Singing (Vocal)   *235*

Appendix II   Glossary   *251*

Part One

Chinese Musical Instruments

Lesson One

Wind Instruments

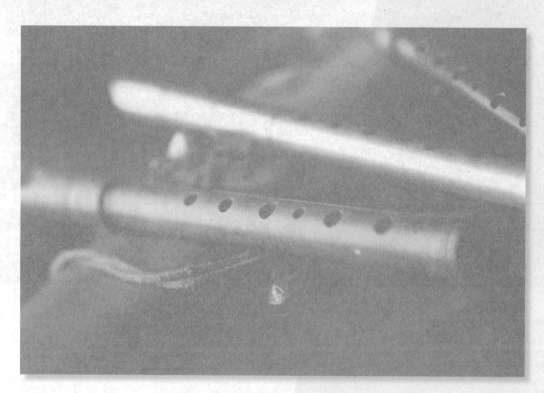

Di

Dizi or *Di*, in Chinese "笛", used to be "涤", meaning to purify. *Di* got its name for the sake of its unique sound —it's the "sound of purification (荡涤之声)". *Dizi* is the oldest and the most representative wind instrument of the Han people.

Now *dizi* is a major Chinese instrument that is widely used in many musical genres: Chinese folk music, Chinese opera, traditional Chinese orchestra, as well as Western orchestra and modern music.

Various of materials are used to make *dizi*. While bamboo has taken the major part, other materials like bone, wood, stone, jade, etc. were also adopted during the ancient times and in other ethnic regions.

I. Construction

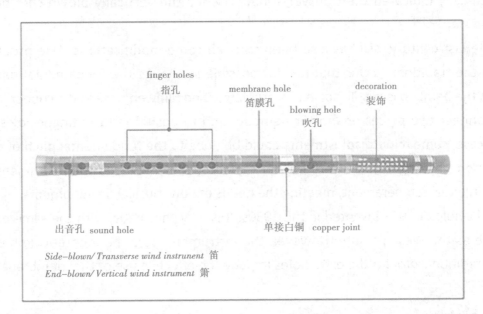

What distinguishes *dizi* from most simple flutes is the membrane hole (dimo), which is between the blowing hole and finger holes and made from the inner skin of bamboo cells. It is extremely thin, almost tissue-like, and traditionally glued by ejiao and garlic juice, which is an art form in itself. In the process of application, some wrinkles are created in the centre of the membrane, making the sound brighter and louder, with a penetrating buzzy timbre and a characteristic nasal quality.

II. Development

Dizi has a fairly long history. The bone flutes, yielded from Jiahu, Henan province are proved coming from the Neolithic age[1] 9,000 years ago. They are so far the earliest transverse wind instruments globally that are still playable today. Those bone flutes were carved with 5 to 8 holes, but mostly with 7 holes.

The bamboo *dizi* is a characteristic instrument of the Han people. As for the origin of using bamboo as the main material, the earliest record is seen in *Records of the Historian*: "Yellow Emperor ordered Linglun to cut bamboos near the creek under the Kunlun Mountains, and make *dizi* out of the bamboo cuts, which was blown to produce sounds like phoenix."[2]

During the long years since *dizi* was invented, the connotation of the name *di* has gone through many changes. Before the Han Dynasty, *di* usually referred to the vertically blown instruments. During the Qin and Han Dynasties, *di* became a general name for both vertical and transverse flutes. Since the Tang Dynasty, *di* has specifically indicated the transverse flutes, while the vertically blown *chi*[3] been called *xiao*.

In the last century, *dizi* has also been through some modifications. One problem with the traditional *dizi* is that it's not possible to change the fundamental tuning once the bamboo is cut. To solve this problem, Zheng Jinwen[4] inserted a copper joint to connect two pieces of shorter bamboo, and re-positioned the finger-holes. In this case, some minute adjustments could be done to the fundamental pitch of *dizi*. Later on, the finger holes were further replaced, which enabled the instrument to play in equal temperament, meeting the needs of new musical developments.

The 11-hole *dizi* was invented in the 1930s. It is fully chromatic, with the same pitch range as the Western flute. However, the instrument lacks the inherent timbre of the traditional *dizi*, for the extra holes impede the effective use of the membrane.

III. Schools

Based on the professional conservatory repertory, the contemporary *dizi* playing is divided into two schools: northern and southern.

The northern school is also called "Beipai", for it is popular in northern China. *Dizi* used in the northern style is called *bangdi*, shorter and higher in pitch, brighter and shriller in sound quality. *Bangdi* is commonly used in Kunqu and Bangzi Opera, as well as some regional genres like Errentai (a song-and-dance duet). The *bangdi* music is featured by fast, rhythmic and virtuosic playing.

The southern school, also known as "Nanpai", is prominent in Kunqu and Jiangnan Sizhu (silk and bamboo music of Jiangnan) of southern China. *Qudi* is the leading melodic instrument in southern folk orchestra, which is longer and with a more mellow, lyrical tone. The music of southern school is slower, with short melodic turns, and appoggiatura or grace notes.

IV. Performing Techniques

Dizi is often played using various techniques, such as:

循环换气	circular breathing/ circle breathing
滑音	portamento
历音	glissando
颤音	trill
吐音	tonguing
双吐	double-tonguing
三吐	triple-tonguing
花舌	flutter-tonguing
打音	lower acciaccature
叠音	upper acciaccature
赠音	note end acciacature
剁音	popped note
泛音	harmonics
飞指	flying finger trill
腹震音	vibrato

Generally speaking, "*tu, hua, duo, hua* (吐、花、剁、滑)" are typical of the northern school, while "*chan, die, zeng, da* (颤、叠、赠、打)" of the southern school.

V. Musical Classics

A Trip to Gu Su 《姑苏行》
The Reed Pipe 《牧笛》

Shepherd Boy	《小放牛》
Flying Partridges	《鹧鸪飞》
Wu Bang Zi	《五梆子》
Busy Transporting Grains by Urging Horses on with Whips	《扬鞭催马运粮忙》

Notes on the text

1. the Neolithic age 新石器时代，大约从距今一万多年前开始，至距今5000多年至2000多年结束。
2. 出自《史记》："黄帝使伶伦伐竹于昆豀、斩而作笛，吹作凤鸣"。
3. *chi* 篪，一种横吹低音竹管乐器，或曰竹埙。
4. Zheng Jinwen 郑觐文 (1872—1935)，江苏江阴人，音乐教育家，民族器乐演奏家，民族乐器改革家。

Terms

blowing hole	/ˈbləʊɪŋ həʊl/	吹孔
finger hole	/ˈfɪŋgər həʊl/	指孔
harmonics/ overtone	/hɑːrˈmɑːnɪks/, /ˈəʊvərtəʊn/	泛音
trill	/trɪl/	颤音
flying finger trill	/ˈflaɪɪŋ ˈfɪŋgər trɪl/	飞指
membrane	/ˈmembreɪn/	笛膜
sound hole	/saʊnd həʊl/	出音孔
circle breathing	/ˈsɜːrkl ˈbriːðɪŋ/	循环换气
portamento	/poʊrtəˈmentoʊ/	滑音
glissando	/glɪˈsændəʊ/	历音
tonguing	/ˈtʌŋɪŋ/	吐音
double-tonguing	/ˈdʌbl ˈtʌŋɪŋ/	双吐
triple-tonguing	/ˈtrɪpl ˈtʌŋɪŋ/	三吐
flutter-tonguing	/ˈflʌtər ˈtʌŋɪŋ/	花舌
note end acciaccature	/nəʊt end ɑːˌtʃɑːkɑːˈtʊərə/	赠音
lower acciaccature	/ˈləʊər ɑːˌtʃɑːkɑːˈtʊərə/	打音
upper acciaccature	/ˈʌpər ɑːˌtʃɑːkɑːˈtʊərə/	叠音
popped note	/pɑːpt nəʊt/	剁音
vibrato	/vɪˈbrɑːtəʊ/	腹震音
transverse	/ˈtrænzvɜːrs/	横向的
vertical	/ˈvɜːrtɪkl/	竖向的

Lesson One
Wind Instruments

Exercises

I. Comprehension questions

1. How did *dizi* get it's name?
2. What are the musical genres that *dizi* could be used?
3. What is the one featured part of *dizi*?
4. Since when has *di* specifically indicated the transverse flutes?
5. Which school often uses *qudi* when performing?

II. Translating useful expressions

1. 笛是中国传统管乐器。
2. 笛是横吹乐器。
3. 笛是汉族最古老、最有代表性的管乐器。
4. It is fully chromatic, with the same pitch range as the Western flutes.
5. Before the Han Dynasty, *di* usually referred to the vertical-blown instruments.
6. *Dizi* used in the northern style is called *bangdi*, shorter and higher in pitch, and brighter and shriller in sound quality.

III. Brainstorm

Please make a research on the similarities and differences between *dizi* and the Western flutes, and report what you have found to the class.

Suona

Suona, also called *laba* or *haidi*, is a double-reed woodwind instrument. As one member of the oboe family in the world, *suona* has a distinctively loud and high-pitched sound, and is used frequently in Chinese traditional music ensembles, particularly those folk *chuida* ensembles performing outdoors. It is an important instrument in the folk music of northern China, particularly the provinces of Shandong and Henan, where it has long been used, in combination with *shengs*, *gongs*, drums and sometimes other instruments, in festivals, weddings and funerals, or for military purposes. It is also common in the ritual music of southeast China, and it is called *guchui* in Taiwan province. Because of its unique timbre, *suona* has become a representative national woodwind instrument in China.

I. Construction

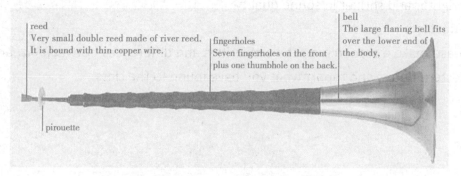

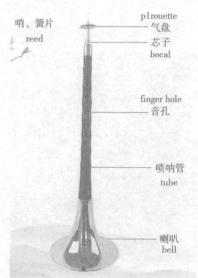

Suona in China has a conical wooden body, a tubular brass or copper bocal to which a small double reed is affixed, and a detachable metal bell at its end.

II. Development

As early as the 3rd century, *suona* was introduced into northwestern China from Central Asia along the Silk Road. The word "*suona*" is derived from the Arabic word "zurna" (oboe), so *suona* possibly has been developed from Central Asian instruments such as the sorna, sunray, or zurna. It was well developed in China by the Yuan Dynasty but it was not mentioned in Chinese literature until the Ming Dynasty (1368–1644).

Since the mid-20th century, the "modernized" versions of *suona* have been developed in China. The mechanical keys and chromatic holes are added to the traditional instruments, as well as the bocal being made contractile, to allow the playing of chromatic notes and to increase the range and stabilize the pitch. *Suona* is made in several sizes and its range is given variously as from one -and- a-half to just over two octaves. There is now a family of the instrument, including piccolo (haidi), soprano *suona* (gaoyin), alto *suona* (zhongyin), tenor *suona* (cizhongyin), and bass *suona* (diyin). These instruments are used in the woodwind sections of modern large Chinese traditional instrument orchestras and in modern music arrangements as well, like in the works of Chinese rock musician Cui Jian, featuring a modernized suona-play in his song "Nothing to My Name".

III. Performing Techniques

Usually *suona* has eight finger holes of which seven finger holes are in the front and one thumb hole is in the back. A *suona* player holds the instrument in front of himself or herself with both hands. He or she completely encloses the reed in his or her mouth without touching it, his or her lips pressed firmly against the pirouette. He or she activates the double reed by using a great deal of air pressure, and the concussion action of the reeds allows rapid bursts of energy into the air column of the instrument to produce sound.

Main playing techniques:
Suona's playing techniques are extremely rich and they can be roughly divided into **oral** and **finger** techniques. They are often used in combination, including:

    legato 连音

single/double/triple-tonguing 单吐、双吐、三吐
flutter-tonguing 花舌音
portamento / slide 滑音
vibrato 颤音
trill 指花
harmonics 泛音
circular breathing 循环换气法

IV. Musical Classics

Hundreds of Birds Paying Homage to the Phoenix	《百鸟朝凤》
Carrying a Wedding Sedan Chair	《抬花轿》
A Song of Harvest Celebration in a Henan Tune	《社庆》
A Piece of Flower Loved by All	《一枝花》
The Salesman Comes to the Mountain Village	《山村来了售货员》
Attachment to the Homeland	《怀乡曲》
Early Spring on the Hanjiang River	《汉江春早》
Song-Dance Duet in Folk Tune	《二人转牌子曲》
Harvesting the Date Fruits	《打枣》
Bangzi Opera Tune in Hebei Province	《河北梆子》

Terms

reed	/riːd/	哨子；簧片
tube	/tuːb/	哨管
bocal	/ˈbɔkal/	芯子
bell	/bel/	喇叭
pirouette	/ˌpɪruˈet/	气盘
finger hole	/ˈfɪŋər həʊl/	指孔
thumb	/θʌm/	大拇指
index finger	/ˈɪndeks ˈfɪŋɡər/	食指
middle finger	/ˈmɪdl ˈfɪŋɡər/	中指
ring finger	/rɪŋ ˈfɪŋɡər/	无名指
pinky finger	/ˈpɪŋki ˈfɪŋɡər/	小指
piccolo	/ˈpɪkələʊ/	短笛
soprano	/səˈprænəʊ/	高音；女高音
alto	/ˈæltəʊ/	中音；女低音
tenor	/ˈtenər/	次中音；男高音

bass	/beɪs/	低音
legato	/lɪˈgɑːtəʊ/	连音
tonguing	/ˈtʌŋɪŋ/	吐音
flutter tonguing	/ˈflʌtərɪŋ ˈtʌŋɪŋ/	花舌音
glide/portamento	/glaɪd/, /ˌpoʊrtəˈmentoʊ/	滑音
vibrato	/vɪˈbrɑːtəʊ/	颤音
harmonics	/hɑːrˈmɑːnɪks/	泛音
circular breathing/circle breathing	/ˈsɜːrkjələr ˈbriːðɪŋ/, /ˈsɜːrkl ˈbriːðɪŋ/	循环换气法
single/double/triple-tonguing	/ˈsɪŋgl ˈtʌŋɪŋ/, /ˈdʌbl ˈtʌŋɪŋ/, /ˈtrɪpl ˈtʌŋɪŋ/	单吐、双吐、三吐

Exercises

I. Comprehension questions

1. Why is *suona* called "Chinese oboe"?
2. When and from where did the ancesters of *suona* enter China?
3. How has *suona* been modernized in the 20th century?
4. Can you name some of the characteristic playing skills of *suona*?
5. Can you describe the structure of *suona* briefly?

II. Translating useful expressions

1. 唢呐是双簧木管乐器。
2. 唢呐声音高亢嘹亮。
3. 唢呐通常有八个指孔。
4. The *suona* family are used in the woodwind sections of modern Chinese instrument orchestras.
5. The "modernized" *suona* adopted mechanical keys similar to those of the European oboe, to allow for the playing of chromatic notes and equal tempered tuning.
6. *Suona* is an important instrument in the folk music of northern China, particularly the provinces of Shandong and Henan, where it has long been used for festivals and military purposes.

III. Brainstorm

Watch a film on *suona* "Song of the Phoenix (百鸟朝凤)", and discuss with the class what kinds of national spirits *suona* has carried.

Sheng

Sheng is a free reed mouth organ, and one of the oldest Chinese reed wind instruments consisting of vertical pipes. It is a polyphonic instrument which uses coupling vibration (耦合振动) to produce sound. The instrument's bamboo pipes, each of a different length, have been likened to a phoenix at rest with its wings closed. "*Sheng*" in Chinese means "to grow and develop", which symbolizes seeds sprouting from the ground.

I. Construction

The Traditional *Sheng*

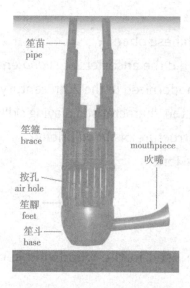

The Keyed *Sheng*

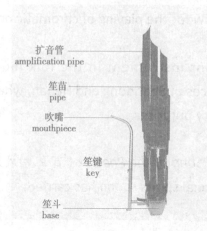

The body of *sheng* has **3 main segments—the first** is the base of *sheng* which includes the metallic base of *sheng* and an attached mouthpiece; **the second** is the pipes, which contains bamboo pipes of different lengths and sizes that extend from the base; **the third** region is the free reeds (笙簧), which are the vibrating reeds found at the bottom of each bamboo pipe.

Sound is produced through vibration of the free reeds causing vibration of the air within the columns in the bamboo pipes. Each pipe has a note window (音窗) with a free beating reed and the window determines the amount of space that air can vibrate within the columns in the bamboo pipes, thus creating a pitch range for the pipes. The accuracy of the pitch is then decided based on the quantity of red wax (红蜡) dotted on the reed. The more wax on the reed, the heavier the reed becomes, causing lower vibrations and a resultant lower pitch. Similarly, the lighter the wax dot, the higher the pitch.

II. Development

Dating back to the Yin Dynasty, the small *sheng* called *he* (和) was mentioned in bone oracle inscriptions. The word " 笙 *sheng* " first appeared in *Shijing* (*Book of Odes*), in the section *Xiaoya Luming* (《小雅·鹿鸣》) "鼓琴吹笙，吹笙鼓簧 , in about 7th century BCE.

Until the Han Dynasty, *yu* (竽) with 22, 23 or 36 reeds and *sheng* with 13, 17 or 19 reeds, had co-existed. *Yu* lost its position in the Song Dynasty, for it mainly being used for ritual purposes.

Numerous historical records have mentioned that *sheng*, as an important ancient music instrument, was popular in the imperial palace and in court processions. *Sheng* was introduced to Japan in the 8th century. *Sheng* was first introduced to Persia through the Silk Road, so Chinese *sheng* has played a positive role in promoting the development of Western musical instruments.

Since the early 20th century, many changes have been made to the structure of *sheng*. The number of pipes has been increased, the size of the pipes thus the air chamber enlarged, and keys and levers were added as well. As a result, the reformed instrument enhanced the sound and volume, expanded its range , allowing it to play harmony and chords.

The rich and dynamic sound qualities of *sheng* make it continue to be popular not only in Chinese traditional venues but also in orchestra.

III. Types

As to the **shapes**, *sheng* can be classified into the **round** *sheng* with a round bottom as a base, and the **rectangular** *sheng* with an oblong base.

Sheng can also be classified into the traditional *sheng* and the improved *sheng* (or the keyed *sheng*). Keyed *shengs* were only developed in the 20th century.

The Traditional *sheng*

The traditional *sheng* players constantly add or remove reed pipes to broaden or reduce their instrument's range at their own direction. The traditional *sheng* cannot bend its pitch with the aid of stronger breath but can do it with the help of fingering techniques. The traditional *shengs* are usually only used for solo repertoire, due to them not being fully chromatic, and are usually held in the player's hands when playing. To enhance its range and volume, after undergoing many changes, the modern traditional *sheng* appeared in the mid-20th century in China. The number of the modern *sheng* pipes increased to 21, and metal tubes were attached to the bamboo pipes to amplify its sound. The other change was the development of the keyed *sheng* which can ease the fingering, and now there are also fully chromatic traditional *shengs*.

The Keyed *sheng*

Keyed *shengs* were created for the Chinese orchestra, as they can play all semitones. Fingering techniques here are not complicated compared to the traditional *sheng*. Current improved *shengs* usually have 32 to 38-reed pipes, however, as these instruments come with levers or keys, it is not possible to bend their pitches. There are four ranges of keyed *sheng*: soprano *sheng*, alto *sheng*, tenor and bass *sheng*. All are chromatic throughout their range, and equal tempered, and they tend to be placed on the player's lap or on a stand while playing.

Among the four keyed *shengs*, due to its mellow timbre, the alto *sheng* plays an important role in modern Chinese orchestra, and it has two main forms in modern Chinese music: *bao sheng* (抱笙) and *pai sheng* (排笙).

In the 21st century, keyboard *shengs*, or *pai shengs* that have a keyboard layout instead of the typical buttons, have emerged. The keyboard *shengs* can vary from 37-reed all the way to 53-reed, covering a variety of ranges from alto to bass.

IV. Performing Techniques

Sheng can be played through sucking and blowing the reeds to create sound. *Sheng* requires a certain amount of blowing force from the performer to be able to vibrate its reeds. It is noted that the higher the note, the greater the amount of force required. Techniques of *sheng* can be split into finger techniques and mouth techniques, but both finger and mouth techniques are not mutually exclusive.

Finger techniques include the pressing of levers, covering of holes and the use of different fingerings to produce different sound effects.

Mouth techniques fulfill two aims—to create different sound textures using breath and to beautify a sound.

There are some stylistic techniques in the sheng performing, such as :

portamento/gliding 滑音 : For the *sheng* performing, gliding is defined as the gradual closing up or opening of the holes in the instrument. Coupled with the control of breathing and fingering techniques, a musician can play an upward glide and a downward glide. Gliding applies to the traditional *sheng* only and cannot be executed on the reformed *shengs*.

flutter tonguing 花舌 : Uses the vibration of the tip of the tongue and the throat to induce continuous columns of air to produce rapid spurts of breath. The lungs are, however, not involved in flutter tongue as the technique only uses the air in the throat.

tremelo by tonguing 呼舌 : Also known as "to and fro air" (来回气), is a difficult technique to master. As the nose breathes, the tongue will move back and forth, creating a constant air column between the reed and the mouth that will make the reed vibrate, hence producing a gentle tidal sound.

Other often used techniques including:
- single tonguing 单吐
- double tonguing 双吐
- triple tonguing 三吐
- soft double tonguing 软双吐
- rapid tonguing 碎吐
- sforzandissimo flutter tonguing 爆花舌
- scale with harmony / parallel notes 和声音阶
- strong vibrato 气震音

chordal appoggiature 抹音
glissando 沥音
portamento 滑音
trills 指颤音
acciaccature 打音
polyphonic 复调

V. Musical Classics

The Phoenix Gets Ready for Flight	《凤凰展翅》
Soldiers Patrolling on the Grassland	《草原巡逻兵》
Spring Dawn in Hainan	《海南春晓》
The Peacock Spreads Its Tail	《孔雀开屏》
The Tune of Jin (Shanxi)	《晋调》
Calling Out to the Phoenix	《唤凤》

Terms

base	/beɪs/	笙斗
mouthpiece	/ˈmaʊθpiːs/	笙嘴
pipe	/paɪp/	笙苗
reed	/riːd/	笙簧
note window	/nəʊt ˈwɪndəʊ/	音窗
red wax	/red wæks/	红蜡
air hole	/er həʊl/	按孔
amplification pipe	/ˌæmplɪfɪˈkeɪʃn paɪp/	共鸣管
single tonguing	/ˈsɪŋɡl ˈtʌŋɪŋ/	单吐
double tonguing	/ˈdʌbl ˈtʌŋɪŋ/	双吐
triple tonguing	/ˈtrɪpl ˈtʌŋɪŋ/	三吐
soft double tonguing	/sɔːft ˈdʌbl ˈtʌŋɪŋ/	软双吐
rapid tonguing	/ˈræpɪd ˈtʌŋɪŋ/	碎吐
flutter tonguing	/ˈflʌtər ˈtʌŋɪŋ/	花舌
sforzandissimo flutter tonguing	/sfɔːtˈsændoʊ ˈflʌtər ˈtʌŋɪŋ/	爆花舌
tremolo by tonguing	/ˈtreməloʊ baɪ ˈtʌŋɪŋ/	呼舌
scale with harmony/ parallel notes	/skeɪl wɪθ ˈhɑːrməni/, /ˈpærəlel noʊts/	和声音阶
strong vibrato	/strɔːŋ vɪˈbrɑːtəʊ/	气震音
chordal appoggiature	/ˈkɔːdəl apɑ(d)ʒjatyːr/	抹音

Lesson One
Wind Instruments

glissando	/glɪˈsændəʊ/	沥音
portamento	/ˌpoʊrtəˈmentoʊ/	滑音
trill	/trɪl/	（指）颤音
acciaccatura	/ɑˌtʃɑkəˈturə/	打音
polyphonic	/ˌpɑːliˈfɑːnɪk/	复调

Exercises

I. Comprehension questions

1. What does "*sheng*" mean in Chinese?
2. Where did *sheng* firstly appear in literature?
3. How was *sheng* introduced to Persia?
4. For what purpose was the keyed *sheng* invented?
5. What are the main parts to construct a *sheng*?

II. Translating useful expressions

1. 笙是使用自由簧的吹奏乐器，是一种复调乐器。
2. 在传统演奏中，笙常常作为笛和唢呐的伴奏乐器。
3. 键笙家族包括：高音笙、中音笙、次中音笙和低音笙。
4. As a result, the reformed instrument enhanced the sound and volume, expanded its range, allowing it to play harmony and chords.
5. The traditional *shengs* are usually only used for solo repertoire, due to them not being fully chromatic, and are usually held in the player's hands when playing.
6. Keyed *shengs* were created for the Chinese orchestra, as they can play all semitones.

III. Brainstorm

Please work with your classmates to make a research on how *sheng* was introduced to the foreign countries and helped the improvement of Western instruments.

Lesson Two

Plucked String Instruments

Lesson Two
Plucked String Instruments

Pipa

Pipa as a traditional Chinese instrument is well-known around the world. It is honored as " the king of the Chinese plucked string instruments". *Pipa* has four strings, a pear-shaped wooden body with a varying number of frets ranging from 12 to 31. The instrument is usually tuned to A2 D3 E3 A3. The term *pipa* refers to two Chinese characters— 琵 (*pi*) and 琶 (*pa*), which symbolize two playing techniques: *pi* is the forward plucking of the string using the index finger and *pa* is the backward plucking of the string with the thumb. Historically *pipa* has been introduced to many other countries. Today *pipa* is one of the most popular Chinese instruments in China.

I. Construction

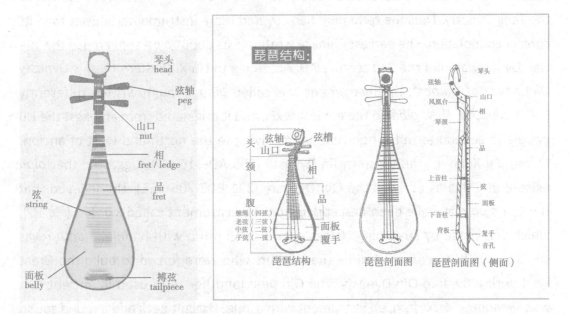

Pipa has three segments—the head, neck and body.

The head section encompasses the nut, tuning pegs and peg box. The neck is usually made of wood, and its pegs can be made of different materials such as ivory, bull's horn or wood. Common designs for the head include peony and lotus flowers, dragons, and phoenixes.

The neck section is covered with frets or ledges (相). Popular materials for frets include wood, ivory, bull's horn and jade.

The body of *pipa* includes frets (品), soundboard, back, fushou (tailpiece) and strings. The fushou, which is usually made of bamboo, holds the strings in place. The strings are usually made of nylon or steel.

The four strings of *pipa* are named differently: the first /thinnest string " 子 zi", the second string " 中 zhong", the third " 老 lao", and the forth/thickest string " 缠 chan".

II. Development

The historical development of *pipa* has been a progressive process from its very beginning and it mainly includes three main evolutionary periods: the Qin *pipa*, the Tang pipa and the post Tang-modern *pipa*.

The Qin *pipa* / Qin-Han *pipa*

There are a lot of arguments about the origin of *pipa* because the word *pipa* was used in ancient texts to describe a variety of plucked chordophones[1] from the Qin to the Tang Dynasty. Thus the term *pipa* had covered more instrumental species than its current connotation. The earliest Chinese written texts about *pipa* appeared in the late Han Dynasty around the 2nd century AD. According to Liu Xi (*Eastern Han*)'s Dynasty *Dictionary of Names*[2], the instrument was called *pipa*, though written differently (琵杷 *pípá*; or 琵把 *pībǎ*) in the earliest texts, and it originated from amongst the Hu people (a general term for non-Han people living to the north and west of ancient China). Fu Xuan[3] of the Western Jin Dynasty (265 AD-316 AD) suggested that *pipa* existed in China as early as the Qin Dynasty (221 BCE-206 BCE). He believed that the Qin *pipa* may have been descended from an instrument called *xiantao* (弦鼗), which was made by stretching strings over a small drum with handle. *Xiantao* was said to have been constructed by the laborers who were forced to build the Great Wall during the late Qin Dynasty. The Qin *pipa* (another term used in ancient text was *Qinhanzi*< 秦汉子 >), an instrument with a long, straight neck and a round sound box, can be played on horseback and it is the predecessor of *ruan*, an instrument named after Ruan Xian[4].

During the Han Dynasty, *pipa*, acquired a number of Chinese symbolisms— the instrument's length of three feet five inches represents the three realms (heaven, earth, and man) and the five elements (metal, wood, water, fire and earth), while the four strings represent the four seasons.

The Tang *pipa*

The pear-shaped *pipa*, which may have been introduced to China from Central Asia, Gandhara, and/or India. From the Southern and Northern Dynasties to the Tang Dynasty, *pipa* had been given various names, such as the Hu *pipa* (胡琵琶) or the bent-neck *pipa* (曲项琵琶). Further fusions and reforms made the pear-shaped pipa increasingly popular through the Sui and Tang Dynasties. By the Song Dynasty, *pipa* referred exclusively to the four-stringed pear-shaped instrument, which were played horizontally with a plectrum. *Pipa* was one of the most popular instruments during the Tang Dynasty, and became a principal musical instrument in the imperial court. Many literary works in Tang refer to *pipa*, for example, Bai Juyi's "*Pipa* Xing/ The Song of a Lute" (《琵琶行》), "Thick strings clatter like splattering rain, /Fine strings murmur like whispered words,/Clattering and murmuring, meshing jumbled sounds,/ Like pearls, big and small, falling on a platter of jade."

The post Tang-modern *pipa* (From Ming and Qing Dynasties to the 20th century)

Because of the influence of neo-Confucian[5] nativism, as the foreign associations, *pipa* lost its position in the imperial court during the Song Dynasty but it was played continually as a folk instrument by the literati. *Pipa* underwent a number of changes over the centuries. By the Ming Dynasty, fingers replaced plectrum as the popular technique for playing, and extra frets were added. The early instrument had 4 frets (相) on the neck, but during the early Ming Dynasty extra bamboo frets (品) were affixed onto the soundboard, increasing the number of frets to around 10 and the short neck of the Tang *pipa* was also lengthened. In the subsequent periods, the number of frets gradually increased, from around 10 to 14 or 16 during the Qing Dynasty, then to 19, 24, 29, and 30 in the 20th century and the 4 wedge-shaped frets on the neck became 6 during the 20th century. In the 1920s and 1930s, the number of frets was increased to 24, based on the 12 tone equal temperament scale, with all the intervals being semitones. The horizontal playing position became the vertical (or near-vertical) position by the Qing Dynasty. During the 1950s, the metal strings replaced the traditional silk ones, which made the sound of the pipa brighter and stronger.

III. Schools

There are a number of different traditions with different styles of playing *pipa* in various regions of China, some of which then developed into schools, which can be divided according to their different forms.

In the narrative traditions where *pipa* is used as an accompaniment to narrative singing, there are the Suzhou Tanci (苏州弹词), Sichuan Qingyin (四川清音), and Northern Quyi (北方曲艺) genres.

Pipa is also an important component of regional chamber ensemble traditions such as Jiangnan Sizhu (江南丝竹), Teochew string music (潮州弦乐) and Nanguan ensemble (南管乐).

There are five main schools associated with the solo tradition, including Wuxi school (无锡派), Pudong school (浦东派), Pinghu school (平湖派), Chongming school (崇明派), Shanghai or Wang school (汪派).

In more recent times, many *pipa* players, especially the younger ones, no longer identify themselves with any specific school. Modern notation systems, new compositions as well as recordings are now widely available and it is no longer important for a *pipa* player to learn from the master of any particular school.

IV. Performing Techniques

The word *pipa* (琵 *pi* and 琶 *pa*) describes how the instrument is played and the sounds it produced. The nylon-wound steel strings are far too strong for human fingernails, so false/ artificial acrylic nails are now used and affixed to the fingertips of the right hand with elastic tape.

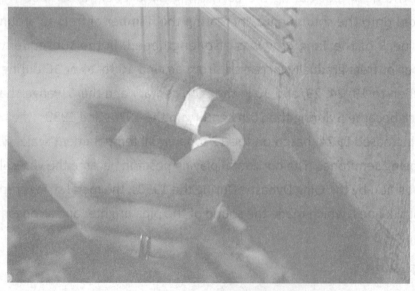

artificial acrylic nails

Right hand

Tan (弹) and **Tiao** (挑) are basic techniques, which involve just the index finger and thumb. Tan is the forefinger plucking outwards from right to left from the player's viewpoint, producing a single sound. Tiao is the thumb plucking the string from left to right from the player's viewpoint, producing a single sound.

Mo (抹) and **Gou** (勾): Plucking in the opposite direction to tan and tiao is called mo (抹) and gou (勾). Mo is the forefinger plucking string inwards from left to right, producing a crisp, thin sound. Gou is the thumb plucking from right to left in a hooking manner, producing a thick but loose sound.

Sao (扫) and **Fu** (拂) are down-stroke brushing and up-stroke brushing. Sao is a rapid strum. The forefinger plucks outwards from right to left, simultaneously plucking all four strings to achieve one sound. Fu is in the reverse direction. The thumb plucks all four strings from left to right to achieve one sound, often used together with sao for a sweeping effect.

Fen (分) and **Zhi** (摭): When two strings are plucked at the same time with the index finger and thumb (the finger and thumb separate in one action), it is called fen (分), the reverse motion is called zhi (摭).

Tremolo (轮指) involves all the fingers and thumb of the right hand, and it is possible to produce the tremolo with just one or more fingers.

Left hand

The left hand techniques are important for the expressiveness of the *pipa* music.

Position (把位): The arrangement of different pitches on the strings.

Vibrato (揉弦): The finger shakes up and down, stopping the string.

Pull-push vibrato (吟弦): The finger shakes left and right, stopping the string.

Harmonics (泛音): It includes natural harmonics and artificial harmonics. Left finger rests slightly on the string at harmonic positions, while the right finger plucks the string to produce sound.

Other common-used techniques includes:

滑音 portamento
打带音 strike and pluck
推拉 pushing and pulling
绞弦 tangle strings
滚奏 short tremolo

V. Musical Classics

The Ambush from All Sides	《十面埋伏》
King Chu Sheds His Armor	《霸王卸甲》
Spring on a Moonlit River	《春江花月夜》
Flute and Drum at Sunset	《夕阳箫鼓》
Snowflakes Falling on the Woods	《飞花点翠》
Whisper from Pipa	《琵琶语》
High Mountains Flowing Waters	《高山流水》
Wang Zhaojun Going Outside the Frontier	《昭君出塞》
Spring Snow	《阳春白雪》
Dragon Boat	《龙船》
Dance of the Yi People	《彝族舞蹈》
Great Waves Sweeping Away Sand	《大浪淘沙》
Heroic Little Sisters of the Grassland	《草原英雄小姐妹》

Notes on the text

1. chordophone 弦鸣乐器
2. *Dictionary of Names* 《释名》，东汉末年刘熙著，是一部专门探求事物名源的佳作。
3. Fu Xuan 傅玄，西晋著名文学家、思想家。
4. Ruan Xian 阮咸，西晋"竹林七贤"之一，善弹琵琶，精通音律。
5. Neo-Confucian 此处指宋明时期新儒学，或曰"宋明理学""程朱理学"。

Terms

nut	/nʌt/	山口
tuning peg	/ˈtuːnɪŋ peg/	弦轴
peg hole/ peg box	/peg hoʊl/, /peg bɑːks/	弦槽
neck	/nek/	琴颈
fret	/fret/	相；品
panel /sound board	/ˈpænl/, /saʊnd bɔːrd/	面板 / 共鸣板
false nail	/fɔːls neɪl/	假指甲
pluck	/plʌk/	弹拨
tremolo	/ˈtremələʊ/	轮指
short tremolo	/ʃɔːrt ˈtremələʊ/	滚奏
pull-push vibrato	/pʊl pʊʃ vɪˈbrɑːtəʊ/	吟弦
vibrato	/vɪˈbrɑːtəʊ/	揉弦

Lesson Two
Plucked String Instruments

portamento	/ˌpoʊrtəˈmentoʊ/	滑音
strike and pluck	/straɪk ən plʌk/	打带音
pushing and pulling	/ˈpʊʃɪŋ ən pʊlɪŋ/	推拉
tangle strings	/ˈtæŋgl strɪŋz/	绞弦
natural harmonics	/ˈnætʃrəl hɑːrˈmɑːnɪks/	自然泛音
artificial harmonics	/ˌɑːrtɪˈfɪʃl hɑːrˈmɑːnɪks/	人工泛音

Exercises

I. Comprehension questions

1. How many strings does a *pipa* have?
2. What is the thinnest string called?
3. What is the Qin-Han *pipa* like?
4. Can you describe the reflections of Chinese symbolisms on *pipa*?
5. Conclude the differences between the Tang *pipa* and modern *pipa*.

II. Translating useful expressions

1. 琵琶是中国拨弦乐器之王。
2. "琵"是指用右手食指往外弹,"琶"是指用右手大拇指往内拨。
3. 直到宋朝,琵琶才专指梨形、四弦的弹拨乐器。
4. *Pipa* has a pear-shaped wooden body with a varying number of frets ranging from 12 to 31.
5. The common designs for the head are peony, lotus flowers, dragons, and phoenixes.
6. From the Southern and Northern Dynasties to the Tang Dynasty, *pipas* were given various names, such as the Hu *pipa* or the bent-neck *pipa*, which was played horizontally with a wooden plectrum.

III. Brainstorm

Please discuss with the classmates: Given the history of *pipa,* what makes it our traditional instrument, regardless of its foreign blood?

Ruan

Ruan is a traditional Chinese plucked string instrument with the history of more than 2,000 years. It is a lute with a fretted neck, a circular body, and four strings. *Ruan* is now most commonly used in Chinese opera and the Chinese orchestra.

I. Construction

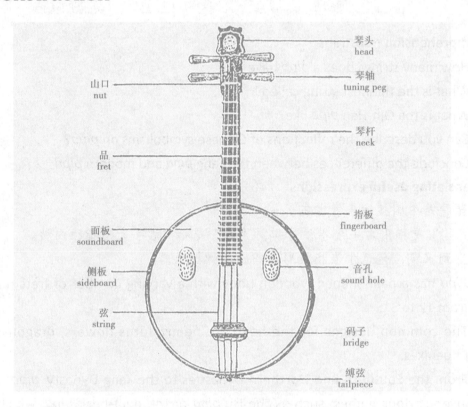

Ruan has three segments—the head, neck and body.

The head is for decoration. Three common decorative patterns are ruyi (good fortune, 如意), peony and dragon. They are usually made of plastic or ivory. Under the lute head is a pegbox with tuning pegs that hold the strings. The pegs are usually made of wood or buffalo horn. The nut is placed between the pegbox and the neck to secure the strings. It is usually made of plastic, buffalo bone or ivory.

The neck connects the lute head and the body. It has a fingerboard and 24 frets. The 24 frets contain 12 semitones on each string, and the frets of a modern *ruan* are set in equal temperament to allow the player to change to any key. The frets are commonly made of ivory or of metal mounted on wood, and the metal frets produce a brighter tone as compared to the ivory frets.

The body, a circular sound box, is made of a combination of front board, back board and side board. There are two sound holes on the front board, through which the sound waves are directed out of the body. Some common shapes for the sound holes are circle, moon, S, and bird. At the lower part of the body, a bridge is placed to support the strings and transmit the vibration. At the bottom of the body is a tailpiece to anchor the strings.

Ruan has four strings, which are numbered from high to low: 1, 2, 3, 4. Its four strings were formerly made of silk but since the 20th century they have been made of steel (flatwound for the lower strings).

II. *Ruan* Family

The modern *ruan* family includes soporano (high pitched *ruan*)[1], alto (small *ruan*)[2], tenor (medium *ruan*)[3], bass (large *ruan*)[4], and contrabass (low pitched *ruan*)[5]. These instruments are constructed almost identically, but in different sizes. In Chinese orchestras, only tenor (medium *ruan*) and bass (large *ruan*) are commonly used, to fill in the tenor and bass section of the plucked string section. Occasionally *gaoyinruan* is used to substitute the high-pitched *liuqin*.

In addition to the plucked *ruan* instruments mentioned above, there also exist a family of bowed string instruments called *laruan* and *dalaruan* (literally "bowed *ruan*" and "large bowed *ruan*").

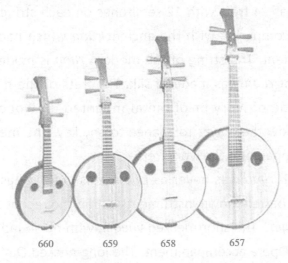

soporano (high pitched ruan), alto (small ruan),
tenor (medium ruan), bass (large ruan) respectively

III. Development

In old Chinese texts from the Han to the Tang Dynasty, the term *pipa* was used as a generic term for a number plucked chordophones, including *ruan*. *Ruan* may have a history of over 2,000 years, and its predecessor may be the Qin *pipa*. However, it is believed that the Qin *pipa* may have been descended from an instrument called *xiantao* (弦鼗), which was constructed by laborers on the Great Wall during the late Qin Dynasty, using strings stretched over a pellet drum. The Qin *pipa* has a long, straight neck with a round sound box in contrast to the pear-shape of *pipa*.

The present name of the Qin *pipa*, which is "*ruan*", was not given until the Tang Dynasty (the 8th century). During the reign of Empress Wu Zetian (about 690 AD-705 AD), a copper instrument that looked like the Qin *pipa* was discovered in an ancient tomb in Sichuan. It had 13 frets and a round sound box. It was believed to be the instrument that Ruan Xian (阮咸) loved to play. Ruan Xian was a scholar in the Eastern Jin Dynasty[6] period (the 3rd century), and one member of the Seven Sages of the Bamboo Grove[7]. Since Ruan Xian was an expert and famous in playing an instrument that looked like the Qin *pipa*, the instrument was named after him when the copper Qin *pipa* was found in a tomb during the Tang Dynasty, but today it is shortened to *ruan*.

Ruan has become a popular instrument in both court music and folk music by the Tang Dynasty. Also during the Tang Dynasty, a *ruanxian* was brought to Japan from China. Today's *ruan* looks like the ancient *ruanxian* because it has made few changes since the 8th century.

The modern *ruan* has 24 frets with 12 semitones on each string, which has greatly widened its range comparing with the ancient *ruan* which had 13 frets, and had no equal temperament. The string of the modern *ruan* is made of steel while the string of the ancient *ruan* was made of silk. The frets of the modern *ruan*, which are commonly made of ivory or of metal mounted on wood, are set in equal temperament to allow the player to change to any key. The metal frets produce a brighter tone as compared to the ivory frets.

Nowadays, although *ruan* was never as popular as *pipa*, it has been divided into several smaller and better-known instruments within the recent few centuries, such as *yueqin*[8] and Qin *qin*[9]. The short-necked *yueqin*, with no sound holes, is now used primarily in Beijing Opera accompaniment. The long-necked Qin *qin* is a member of both Cantonese[10] and Chaozhou[11] ensembles.

IV. Performing Techniques

Ruan can be played by using a plectrum, or by using a set of two or five fingernails (or acrylic nails) that are affixed to the fingers with adhesive tape. Many *ruan* players use plectrums, though there are some schools which teach the fingernail technique, similar to that of *pipa*. Plectrums produce a louder and clearer tone, while fingernails allow the performance of polyphonic solo music. *Ruan* produces a mellow tone.

Right hand techniques:

  down & up stroke 弹挑
  down-stroke brushing 扫
  up-stroke brushing 拂
  striking 打
  tremolo 轮
  double Strings 双音 / 双弹
  mute 伏 / 煞音

Left hand techniques:

The *ruan* player uses 4 fingers of the left hand to play the notes, just like the guitarist. The fingers are numbered as 1 (forefinger), 2 (middle finger), 3 (ring finger), and 4 (pinky).

  vibrato 吟 / 揉
  portamento 滑音
  trill 颤音
  position 把位
  thumb position 大指把位

Both hands techniques:

  slap 拍
  harmonics 泛音

V. *Ruan* and *Pipa*

Similarity

Both of them belong to the plucked sting instrument.

Differences

Ruan has a 24-fretted neck and a circular body while *pipa* has a pear-shaped wooden body with a varying number of frets ranging from 12 to 31.

Ruan is an indigenous Chinese instrument while the pear-shaped *pipa* was not brought to China until the Northern Wei period (386 AD-534 AD) when ancient China traded with the Western countries through the Silk Road[12].

As the king of the Chinese plucked string instruments, *pipa* is one of the most popular Chinese instruments in China and has been introduced to many other countries. Nowadays, although *ruan* was never as popular as *pipa*, *ruan* has been divided into several smaller and better-known instruments within the recent few centuries, such as *yueqin* and Qin *qin*.

VI. Musical Classics

Red All Over the River	《满江红》
Water Lilies	《睡莲》
The Night of the Torch Festival	《火把节之夜》
The Mountain Charm	《山韵》
Love of the Han Pipa	《汉琵琶情》
Camel Bells Along the Silk Road	《丝路驼铃》
Narration of Yuguan	《玉关引》
Between the Sky and the Land	《天地之间》
The Water-Sprinkling Festival	《泼水节》
Phoenix Flowers Are in Blossom	《凤凰花开》
Tone Poem Beyond the Great Wall	《塞外音诗》
Reminiscences of Yunnan	《云南回忆》
The Turned Over Curtain	《倒垂帘》
Zhong Ruan Rock 'N' Roll/ The Man's Knife	《中阮摇滚/男人的刀》

Notes on the text

1. soprano: *gaoyinruan* 高音阮 , lit. "high pitched *ruan*"

Lesson Two
Plucked String Instruments

2. alto: *xiaoruan* 小阮, lit. "small *ruan*"
3. tenor: *zhongruan* 中阮, lit. "medium *ruan*"
4. bass: *daruan* 大阮, lit. "large *ruan*"
5. contrabass: *diyinruan* 低音阮, lit. "low pitched *ruan*"
6. Eastern Jin Dynasty 东晋（公元 317 年—公元 420 年），是由西晋皇族司马睿南迁后建立起来的王朝。在中国历史上，东西两晋南北朝时期，社会较为动荡、混乱，但也是文化交流、民族融合的时期。
7. the Seven Sages of the Bamboo Grove 竹林七贤 (Ruan Xian and other six scholars disliked the corruption government, so they gathered in a bamboo grove in Shanyang. They drank, wrote poems, played music and enjoyed the simple life. The group was known as the Seven Sages of the Bamboo Grove.)
8. *yueqin*: "moon" lute, 月琴
9. Qin *qin*: Qin [Dynasty] lute, 秦琴
10. Cantonese 广东
11. Chaozhou 潮州，位于广东省内。
12. the Silk Road 丝绸之路，起源于西汉（公元前 202 年—公元 8 年）。

Terms

lute head	/luːt hed/	琴头
tuning peg	/ˈtuːnɪŋ peg/	琴轴
neck	/nek/	琴杆
fingerboard	/ˈfɪŋɡərbɔːrd/	指板
nut	/nʌt/	山口（西方乐器：琴枕）
fret	/fret/	品
board	/bɔːrd/	面板
side board	/saɪd bɔːrd/	侧板
sound hole	/saʊnd həʊl/	音孔
bridge	/brɪdʒ/	琴马
tailpiece	/ˈteɪlpiːs/	缚弦；系弦板
pegbox	/ˈpegbɒks/	弦轴箱
string	/strɪŋ/	弦
plectrum/ single pick	/ˈplektrəm/, /ˈsɪŋɡl pɪk/	琴拨/弹片
acrylic/ artificial nail	/əˈkrɪlɪk neɪl/, /ˌɑːrtɪˈfɪʃl neɪl/	假指甲
ensemble	/ɑːnˈsɑːmbl/	重奏
down & up stroke	/daʊn ən ʌp strəʊk/	弹挑
down-stroke brushing	/daʊn strəʊk ˈbrʌʃɪŋ/	扫

up-stroke brushing	/ʌp stroʊk ˈbrʌʃɪŋ/	拂
striking	/ˈstraɪkɪŋ/	打
tremolo	/ˈtremələʊ/	轮
double strings	/ˈdʌbl strɪŋz/	双音 / 双弹
mute	/mjuːt/	伏 / 煞音
vibrato	/vɪˈbrɑːtoʊ/	吟 / 揉
portamento	/ˌpoʊrtəˈmentoʊ/	滑音
trill	/trɪl/	颤音
position	/pəˈzɪʃn/	把位
thumb position	/θʌm pəˈzɪʃn/	大指把位
slap	/slæp/	拍
harmonics	/hɑːrˈmɑːnɪks/	泛音

Exercises

I. Comprehension questions

1. Is *ruan* a plucked instrument?
2. Why does *ruan* name after Ruan Xian?
3. What is the function of nut in a *ruan*?
4. What are the main right hand performing techniques when playing *ruan*?
5. What aspects have the differences between *pipa* and *ruan*?

II. Translating useful expressions

1. 阮是一种有音箱、圆形、24 品和四弦的弹拨乐器。
2. 阮用琴拨或一套假指甲演奏。
3. 中阮和大阮常用于京剧和中国管弦乐团，以丰满弹拨乐器组的中低音声部。
4. In old Chinese texts from the Han to the Tang Dynasty, the term *pipa* was used as a generic term for a number plucked chordophones, including *ruan*.
5. The 24 frets have 12 semitones on each string, and the frets of modern *ruan* are set in equal temperament to allow the player to change to any key.
6. The performing technique of "Mute" (or "Wu" or "Sha") is to use the palm to cover the strings to stop their vibration after plucking.

III. Brainstorm

The frets on all Chinese lutes are high so that the fingers never touch the actual body—distinctively different from Western fretted instruments. This allows for a greater control over timbre and intonation than their Western counterparts, but makes chordal playing more difficult. Please take *ruan* (or *pipa*) and the guitar as examples, by comparing and researching their differences in construction, timbre, intonation and playing techniques, to make clear the differences between the Chinese fretted instruments and Western fretted ones.

Guqin

Guqin, which was originally called "*qin* (琴)", is a plucked seven-string Chinese musical instrument. Later, as many Western instruments were imported, the term "*qin*" had been applied to many other musical instruments as well. People added the character "*gu*", meaning "the ancient", to distinguish it from other musical instruments such as *yangqin*, *huqin* and the piano.

Guqin has 7 strings with a range of about 4 octaves. Its open strings are tuned in the bass register. The lowest pitch is about C2. The sounds are produced by plucking open strings, stopping strings, and making harmonics.

Guqin is mainly used in solo context, for its quietness of tone. However, it can be accompanied by other *qin*, or *xiao*, or singing. *Xiao* playing with *qin* is known as *qinxiao* (琴箫), which has a narrower range. And the literatures that discuss *qin* lore, *qin* theory, *qin* music, etc. are called *qinpu* (*qin* tablature collections).

Guqin is considered as an indispensable part of Chinese elite culture. It ranks the top of the *Four Arts*: *qin*, Go, calligraphy and painting. In the ancient times, decent noblemen and scholars would be able to play it. The instrument even defines the essence of aestheticism and philosophy in China. It is sometimes called by Chinese as "the father of Chinese music" or "the instrument of the sages", as well as the most frequently shown-up subject matter in Chinese literature and has been interweaving with the history of the country. In 2003, *guqin* was proclaimed as an Intangible Heritage of Humanity by UNESCO[1].

I. Construction

There are a lot of symbols in *guqin*. In Chinese measurement, the entire length of *guqin* is 3 chi 6 cun 5 fen, representing the 365 days in a year. The top surface is round and convex, representing the sky. The bottom surface is flat, representing the earth.

Body: The design of its body is modeled on the body of phoenix, which is a legendary bird of wonder. Also, corresponding to the human body, the body of *guqin* has head, neck, shoulder, waist, feet, etc. The body of *guqin* is a long, narrow, hollow box made from two pieces of wooden board, and the top board is carved into an arch while the bottom is flat. For the top board, soft wood is usually used such as tung, while the wood for the bottom board is hard catalpa or fir. The surface of the box is covered with a special layer (about 1 mm) of roughcast, which is a mixture of deer horn powder (or bone powder or tile powder) and raw lacquer, and there are several layers of raw lacquer along the top of the roughcast for polishing.

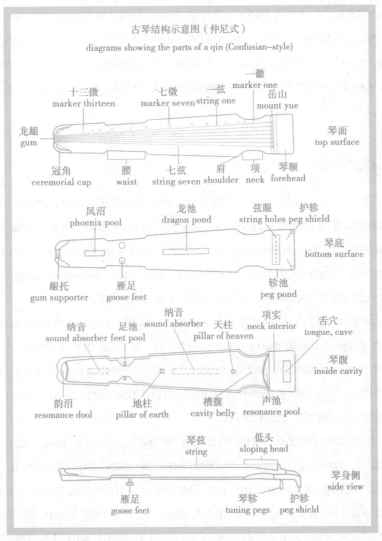

Dragon pool/Phoenix pool: There are two sound holes in the bottom board, and, directly above, on the inner side of the top board there are two protrusions (纳音) with the same shape as the sound holes.

String: It has 7 strings, the first 5 presenting the 5 elements of metal, wood, water, fire and the earth in pentatonic scale. Until the 1960s, the strings of *guqin* were always made of various thicknesses of twisted silk. The top thicker strings from one to four are further wrapped in a thin silk thread, coiled around the core to make it smoother. To chase the longer durability and louder tone, modern nylon-wound steel strings are used nowadays. But some players still prefer the silk strings because of its refinement of tone. The strings are stretched over the bridges, across the surface board, over the dragon gums to the back board, where the end is wrapped around the goose feet.

Hui: There are 13 huis on the surface to indicate the sound position, which represent the 13 months of Chinese lunar year. The huis are filled in *guqin* with shells, gold, silver, jade or pearl. The biggest hui (emblem) represents the lunar month. The huis indicate the positions of overtones and also function as a reference for stopped strings.

II. Development

According to the legend, *qin* has a history of about 5,000 years. The Chinese believe that *qin* was originally invented by Fuxi[2], a tribe leader and hero of the mythological period in ancient China. Legendary figures of China's prehistory— Shengnong[3] and Huang Di[4]—were involved in the creation of "*qin*" too. The exact origin of *qin* is still a continuing subject of debate over the past few decades.

The ancient form of *qin* was short—almost a third of the size of a modern *qin*—and probably only played using open strings. This is because the surface of these early *qins* were not smooth like the modern *qin*. It had engravings on it which would make sliding impossible. The strings were far away from the surface and did not mark the harmonic positions for the players too.

Chinese tradition says *qin* originally had five strings, and later two more were added about 1000 BCE, making it seven. In the Zhou Dynasty, the king Zhou Wen Wang[5] added the 6th string to mourn his son, Bo Yikao[6], after the death of this son. So this string is called Wen string. Later, his successor, the king Zhou Wu Wang[7], added the 7th string to encourage his soldiers to fight bravely with the troops of Shang Kingdom. The 7th string is Wu string.

Ji Kang[8] (223 AD-262 AD) described qin in his poetical essay "Rhapsody on the Qin"—"*Qin* Fu" (《琴赋》) that the basic shape of *qin* was most likely standardized around the late Han Dynasty, which had 7 strings, a small soundbox, with no "hui". To the Weijin-Southern and Northern Dynasties, *qin* became increasingly prevalent among literati, who not only played *qin*, but composed a lot of pieces. The Seven Sages of the Bamboo Grove (竹林七贤) were the representatives. Until the Sui and Tang Dynasties, the popularity decreased a little because of the introduction of the Hu music. The shape of the Tang *qin* was comparatively fat and round. During the Song Dynasty, people's favor for *qin* achieved the prime. Its shape changed to be long and flat, with the entire length of 128cm, larger than the Tang *qin*. Since the late Qing Dynasty, the art of *qin* has declined.

The most famous surviving Tang *qin* was named "Jiuxiao Huanpei(九霄环佩)[9]", attributed to the famous late Tang Dynasty *qin* maker Lei Wei (雷威). It is kept in the Palace Museum inside the Forbidden City in Beijing.

III. Performing Techniques

According to the book *Cunjian Guqin Qupu Jilan* (《存见古琴曲谱辑览》), there are around 1070 different finger techniques, thus *qin* is the instrument with the most playing techniques in both Chinese and Western instrument families.

There are 3 category sounds produced by *guqin*. Harmonics (泛音) represent the sound from the heaven, by plucking a string with the right hand and gently tapping specific note position with the left hand. Open strings (散音) represent the sound from the earth, by plucking a free string with the right fingers. Stopped strings (按音) represent the human beings, by stopping a string with the left finger, striking it with the right hand, meanwhile sliding the left hand up and down to vary the note.

There are eight basic playing techniques:

Tiao (挑): Using the index finger to strum the string outwards. When playing, put your thumb tip against your index finger and slightly bend the index finger to strum the string.

Gou (勾): Using the middle finger to pluck the string inwards. Slightly bend the middle finger, and put more strength on the fingertip. After you pluck the string, stay your middle finger on the next string and not move away yet.

Pi (劈): Using the thumb to strum the string outwards.

Bo (拨): Cupping the index, middle and ring fingers together and plucking the string(s) (one or two) strongly inwards and fast to the left.

Mo (抹): Using the index finger to pluck the string inwards.

Ti (剔): Using the middle finger to strum the string outwards.

Da (打): Using the ring finger to pluck the string inwards.

Zhai (摘): Using the ring finger to pluck the string outwards.

Other often-used techniques are:

Gun 滚 / **Fu** 拂 glissando
Chuo 绰 up-sliding
Zhu 注 down-sliding
Jinfu 进复 up-back sliding
Tuifu 退复 down-back sliding
Yin 吟 / **Rou** 揉 vibrato
Lun 轮 tremolo

IV. Schools

The *qin* schools are known as *qin* pai, which were generally formed around areas. The first *qin* school— Zhe school (浙派) appeared in the late Southern Song Dynasty. During the Ming and Qing Dynasties, various schools emerged, among which differences are often in the interpretation of the music. The traditional schools are: Zhe school (浙派), Yushan school (虞山派), Zhucheng school (诸城派), Mei'an school (梅庵派), Guangling school (广陵派), Jiuyi school (九嶷派), Pucheng school (浦城派), Chuan school (泛川派), Lingnan school (岭南派). Generally speaking, northern schools tend to be more vigorous in techniques than southern schools.

V. Musical Classics

Wild Geese Flock to Sandy Shores	《平沙落雁》
Plum Blossom in Three Movements	《梅花三弄》
Mist and Clouds over the Xiao and Xiang	《潇湘水云》
Spring Snow	《阳春白雪》
High Mountains and Flowing Waters	《高山流水》
Farewell at the Yangguan Pass	《阳关三叠》
Guang Ling Verse	《广陵散》
A Conversation Between a Fisherman and a Woodcutter	《渔樵问答》
Song of the Homebound Fishermen	《渔舟晚唱》

✎ Notes on the text

1. Intangible Heritage of Humanity by UNESCO 联合国教科文组织非物质文化遗产
2. **Fuxi** 伏羲是中国古代传说中的中华民族人文始祖，是中国古籍中记载的最早的王，也是中国医药鼻祖之一。
3. **Shengnong** 神农，即炎帝，是中国上古时期姜姓部落的首领尊称。
4. **Huang Di** 黄帝，中国古代部落联盟首领，五帝之首，被尊祀为"人文始祖"。
5. **Zhou Wen Wang** 周文王，周朝开国君主（约公元前1152年—公元前1056年）。
6. **Bo Yikao** 伯邑考，周文王姬昌嫡长子。
7. **Zhou Wu Wang** 周武王（？—约公元前1043年）
8. **Ji Kang** 嵇康，三国时期曹魏思想家、音乐家、文学家。
9. **Jiuxiao Huanpei** 九霄环佩琴，唐，伏羲式。

Terms

dragon's gum/nut	/ˈdræɡənz ɡʌm/, /ˈdræɡənz nʌt/	龙龈
ceremonial cap	/ˌserɪˈməʊniəl kæp/	冠角；焦尾
marker	/ˈmɑːrkər/	徽位
waist	/weɪst/	龙腰
string	/strɪŋ/	琴弦
shoulder	/ˈʃəʊldər/	仙人肩
string hole	/strɪŋ həʊl/	弦眼
mount yue / bridge	/brɪdʒ/	岳山
cheng lu		承露
neck	/nek/	凤项
head	/hed/	凤头
dragon pool	/ˈdræɡən puːl/	龙池
phoenix pool	/ˈfiːnɪks puːl/	凤池
gum supporter	/ɡʌm səˈpɔːrtər/	龈托
tuning peg	/ˈtuːnɪŋ peɡ/	琴轴
peg shield/ protector	/peɡ ʃiːld/, /peɡ prəˈtektər/	护枕
phoenix forehead	/ˈfiːnɪks ˈfɔːrhəd/	凤额
vibrato	/vɪˈbrɑːtəʊ/	吟；揉
open tone	/ˈəʊpən təʊn/	散音
harmonics/ overtone	/hɑrˈmɑnɪks/, /ˈoʊvərˌtoʊn/	泛音
stopped tone	/stɑːpt təʊn/	按音
glissando	/ɡlɪˈsændəʊ/	滚；拂
up-sliding	/ʌp ˈslaɪdɪŋ/	绰
down-sliding	/daʊn ˈslaɪdɪŋ/	注
up-back sliding	/ʌp bæk ˈslaɪdɪŋ/	进复
protrusions	/prəˈtruːʒənz/	纳音

Exercises

I. Comprehension questions

1. How did *guqin* get its name?
2. How does *guqin* produce sounds?
3. What do the top surface and bottom surface of *guqin* symbolize respectively?
4. What does the biggest hui represent?
5. What are the three category sounds made by *guqin*?

II. Translating useful expressions

1. 古琴是中国传统拨弦乐器。
2. 古琴音色深沉悠远。
3. 拇指向外扫弦。
4. The top thicker strings from one to four are further wrapped in a thin silk thread, coiled around the core to make it smoother.
5. The huis indicate the positions of overtones and also function as a reference for stopped strings.
6. The top surface is round and convex, representing the sky. The bottom surface is flat, representing the earth.

III. Brainstorm

Please read the poem *A Poem about Qin* by Su Shi, and discuss with your classmates: What are the essential elements in playing an instrument?

琴诗

— 苏轼

若言琴上有琴声，放在匣中何不鸣？

若言声在指头上，何不于君指上听？

Guzheng

Guzheng, is also called *zheng*, or *qinzheng* for its first spreading in Qin area (now Shaanxi). It is a Chinese traditional pluck-stringed instrument, nicknamed Chinese zither. Commonly, the modern *guzheng* has a large resonant soundbox of about 1.6 meters long, with 21, 25 or 26 strings. It is tuned in a major pentatonic scale: do, re, mi, so and la, fa and ti being produced by bending the strings. *Guzheng* is generally decorated by various art forms, such as mother-of-pearl inlays, calligraphy, carved art, shell or jade carving and cloisonne.

As an ancient Eastern ethnic instrument, *guzheng* is the forerunner of many plucked string instruments in Asian countries, including the Japanese koto[1], the Korean gayageum[2] and the Mongolian yatga[3].

I. Construction

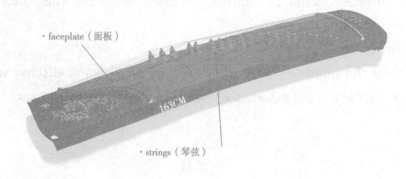

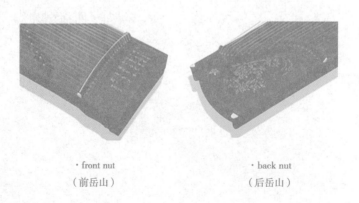

Lesson Two
Plucked String Instruments

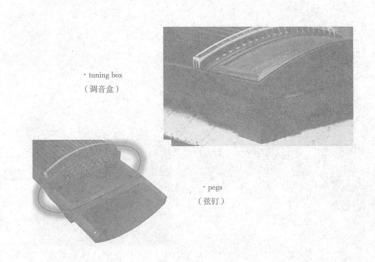

- tuning box
（调音盒）

- pegs
（弦钉）

- holes for
penetrating stings
（穿弦孔）

- goose pillars / bridges
（雁柱/筝码）

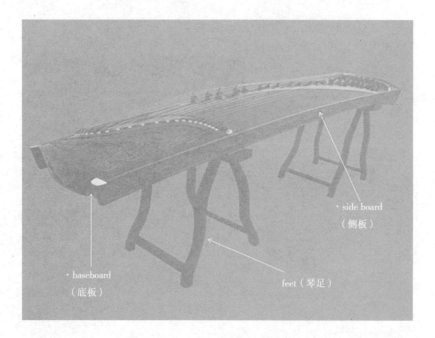
· side board
（侧板）
· baseboard
（底板）
feet（琴足）

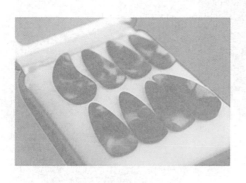

· finger picks/ false nails
（义甲）

II. Development

There are many accounts for the origin of *guzheng*, but basically three major ones. The first saying goes " Meng Tian[4] creating *zheng* (蒙恬造筝)", saying it was invented by Meng Tian, a general of the Qin Dynasty. The second account comes from *The Dictionary of Names* by Liu Xi in the Eastern Han Dynasty[5], recording " when playing fast and hard, the instrument produces sounds like knocking the metal (筝筝然)". So it was named after the sound of the instrument. The third one being " breaking *se* into *zheng*", it is said that two people fought over a 25-string *se* (瑟)[6],

breaking it into halves, each with 12/13 strings, naming each half *zheng* (筝) after the fight (争).

During its long history, *guzheng* has undergone many changes. It has existed since the Warring States Period and became prominent in the Qin State (now Shaanxi province), so it is also called *qinzheng* (秦筝) . The early *zheng* had only 5 strings and a bamboo body (筑身). In the Three Kingdoms Period, the body was strung with 12 strings, made of wood instead of bamboo, which had a bigger soundbox like *se* (瑟). During the Tang and Song Dynasties, *zheng* reached its prime time, and became the most commonly-played instrument, when the strings were added to 13. During the Dynasties of Yuan, Ming and Qing , *zheng* evoled in a slow pace. However, the strings were added to 14 in the Yuan Dynasty, and 15 in the Ming Dynasty, 16 in the late Qing Dynasty. Nowadays it commonly has 21, 25 or 26 strings, but some traditional musicians still use the 16-string, especially along the southeastern coastal provinces.

The strings used to be made of silk, then transitioned to bare wire such as brass in the Qin Dynasty. Nowadays multi-materials, e.g. steel coated in nylon, are widely chosen for modern strings, to increase the volume of *guzheng*, as well as maintain an acceptable timbre.

III. Schools

In terms of playing techniques and repertoires, the traditional *guzheng* playing varies in northern and southern regions, thus has formed northern style and southern style.

Northern style includes Shandong school and Henan school.

Shandong *zheng* is closely conected with Shandong Qinshu. Its melody lines are full of dramatic ups and downs, while music light and refreshing. Glissando are always on beat. Representative songs include *High Mountain and Flowing Waters* (Shandong version) and *Autumn Moon over the Han Palace*.

Henan school is known for its simplicity and elegance. Its songs are often fiery and shown with more slide descending notes than those of Shandong. As for the techniques, it frequently uses left hand slides and vibrato, and the tremolo is made with the thumb. Its representative songs include *High Mountain and Flowing Waters* (Henan version)" and *Going Upstairs*.

Southern style includes Chaozhou and Hakka (Kejia) school, and Zhejiang school.

Chaozhou and Hakka school features embellished melodies. Down sliding notes are used even less, and glissando is free rhythm. Beats are irregular, and alternate between hard and soft taps on the strings. The representative songs include *Jackdaw Plays with Water* and *Lotus Emerging from Water*.

Zhejiang school adopts techniques similar to that of *pipa*, using frequent tremolo with left-hand glissando. It also has some characteristic techniques, such as sidian, where 16 notes are quickly played by thumb, first and second fingers. The representative songs include *The General's Command*.

IV. Performing Techniques

Traditional style: Use the right hand to pluck notes; use the left hand to Zuo Yun (rhyme skills) : add ornamentation such as pitch slides and vibrato by pressing the strings to the left of the movable bridges.

Modern style: Require more use of left hand to provide harmony and bass note. This gives *guzheng* a more flexible musical range, but also prevents the subtle ornamentation provided by the left hand.

Basic performing techniques:

勾 **gou**: Pluck inward with middle finger.

剔 **ti**: Pluck outward with middle finger.

抹 **mo**: Pluck inward with index finger.

挑 **tiao**: Pluck outward with index finger.

托 **tuo**: Pluck outward with thumb.

劈 **pi**: Pluck inward with thumb.

摇指 **yaozhi**: Pluck the strings by shaking rapidly backward and forward with thumb and index finger nipping together.

大撮 **dacuo**: Pluck an octave with thumb and middle finger picking at the same time

小撮 **xiaocuo**: Pluck by thumb and index finger picking at the same time, usually with an interval of fifth, fourth or third.

刮奏 : Glissando

流水 : Flowing water / repeating glissando

上滑音 : Up portamento

下滑音 : Down portamento

颤音：Vibrato
轮指：Tremolo
扫弦：Brushing
弦外音：Non-pitch tone

V. Musical Classics

Song of the Homebound Fishermen	《渔歌唱晚》
Lotus Emerging from Water	《出水莲》
High Mountains and Flowing Waters	《高山流水》
Lin Chong Flees in the Night	《林冲夜奔》
Dance Music of the Dong People	《侗族舞曲》
Autumn Moon over the Han Palace	《汉宫秋月》
Jackdaws Playing in the Water	《寒鸦戏水》
Fisherman's Song of the East Sea	《东海渔歌》
Drum-Shooting on Xiang Mountain	《香山射鼓》
Battle Against Typhoon	《战台风》

Notes on the text

1. koto 十三弦筝（日本筝）
2. gayageum 伽倻琴（韩国筝）
3. yatga 雅托葛（蒙古筝）
4. Meng Tian 蒙恬（？—公元前 210 年），齐国蒙山（今山东省临沂市蒙阴县联城乡边家城子村）人，秦朝时期名将。
5. *The Dictionary of Names* by Liu Xi in the Eastern Han Dynasty 东汉刘熙所作《释名》
6. *se* 瑟，汉族古弹拨乐器，二十五根弦。

Terms

face plate	/feɪs pleɪt/	面板
side board	/saɪd bɔːrd/	侧板
baseboard	/ˈbeɪsbɔːrd/	底板
foot	/fʊt/	琴足
string	/strɪŋ/	弦
front nut	/frʌnt nʌt/	前岳山
tuning box	/ˈtuːnɪŋ bɑːks/	调音盒
tuning pin	/ˈtuːnɪŋ pɪn/	弦钉

bridge	/brɪdʒ/	雁柱
false nail/ finger pick	/fɔːls neɪl/, /ˈfɪŋər pɪk/	义甲
pentatonic scale	/ˌpentəˈtɑːnɪk skeɪl/	五声音阶
back nut	/bæk nʌt/	后岳山
hole for penetrating string	/hoʊl fɔːr ˈpenətreɪtɪŋ strɪŋ/	穿弦孔
sliding	/ˈslaɪdɪŋ/	滑音
tremolo	/ˈtreməloʊ/	轮指
pentatonic scale	/ˌpentəˈtɑːnɪk skeɪl/	五声音阶
vibrato	/vɪˈbrɑːtəʊ/	颤音
tremolo	/ˈtremələʊ/	轮指
glissando	/glɪˈsændəʊ/	刮奏
flowing water/ repeating glissando	/ˈfloʊɪŋ ˈwɔːtər/, /rɪˈpiːtɪŋ glɪˈsændəʊ/	流水
up portamento	/ʌp ˌpoʊrtəˈmentoʊ/	上滑音
down portamento	/daʊn ˌpoʊrtəˈmentoʊ/	下滑音
harmonics	/hɑrˈmɑnɪks/	泛音
brushing	/ˈbrʌʃɪŋ/	扫弦
non-pitch tone	/ˈnoʊn pɪtʃ toʊn/	弦外音

Exercises

I. Comprehension questions

1. What category of instruments does *guzheng* belong to?
2. How is *guzheng* tuned?
3. What are the general 3 statements accounting for the invention of *guzheng*?
4. What are the common materials to make strings nowadays?
5. What is the unique playing technique in Zhejiang school?

II. Translating useful expressions

1. 古筝属中国传统拨弦乐器。
2. 古筝是五声音阶定弦。
3. 古筝在战国时期的秦国境内开始广泛流行，故又名"秦筝"。
4. The use of multi-materials to make strings nowadays increased the instrument's volume, while maintaining the acceptable timbre.

Lesson Two
Plucked String Instruments

5. In terms of playing techniques and repertoires, the traditional *guzheng* playing varies in northern and southern regions, thus has formed northern style and southern style.

6. Henan school is known for its simplicity and elegance.

III. Brainstorm

Please make a research on the several accounts for the invention of *guzheng*, and then choose one account to tell a story, with reasonable imagination.

Konghou

Konghou is a Chinese multi-stringed, plucked instrument of the harp family, also called the "Chinese harp". The ancient *konghou* has three shapes: the horizontal *konghou*, the vertical *konghou* and the phoenix-headed *konghou*. *Konghou* has been handed down for more than two thousand years. It was used to play *yayue* (court music) in the Kingdom of Chu. During the Han Dynasty (202 BCE-220 AD), as "the voice of Chinese nation", *konghou* was included in the imperial music of *qingshangyue* (a music genre). Beginning in the Sui Dynasty (581 AD-618 AD), *konghou* was also used in *yanyue* (banquet music). *Konghou* playing was most popular in the Sui and Tang Dynasties, and it was generally played in rites and ceremonies and gradually prevailed among the ordinary people. Until the Ming Dynasty, it became extinct. In the 20th century, after undergoing many reforms, *konghou* has been revived as a double bridge harp. The modern version of the instrument does not resemble the ancient one, but its shape is similar to Western concert harps.

I. Construction

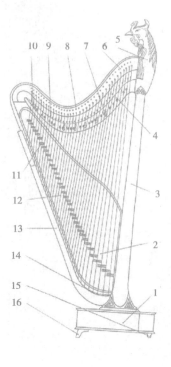

1. base 底座
2. sound-box 共鸣箱
3. column/pillar 立柱
4. modulation system 转调系统
5. scroll 琴头
6. S-shaped panel S 形面板
7. modulation handle 转调手柄
8. tuning pin 琴钉
9. nut 岳山
10. movable nut 活动岳山
11. bridge 琴马
12. string 琴弦
13. strings cover 护弦罩
14. tail pin 挂弦钉
15. movable roller 活动滚轮
16. feet 底脚

The strings of a modern *konghou* are folded over to make two rows, which enable players to use advanced playing techniques such as vibrato and bending tones. The paired strings on opposite sides of the instrument are tuned to the same note, starting from a tuning peg and beyond the playing area travel over two bridges on opposite sides of the instrument, then fixed at the far end to opposite sides of a freely moving lever so that depressing one of the string pairs raises the pitch of the other. The two rows of strings also make it suitable for playing swift rhythms and overtones.

II. Development

The horizontal *konghou*

The horizontal *konghou* was also called "*kanhou* 坎候", and it was first mentioned in written texts in the Spring and Autumn Period. At that time the horizontal *konghou* had 5 strings and 10 frets, and was plucked or struck with a slice of bamboo. The horizontal *konghou* was most prevalent from the Han Dynasty to the Sui and Tang Dynasties, for example, in the Yuefu folk songs in the Han Dynasty, *Peacocks Fly from the East to the South* says that Liu Lanzhi, an ordinary woman, "could weave at the age of 13 and do sewing at the age of 14. At 15 she could play *konghou* and could recite poems at 16".[1] The horizontal *konghou* was not only popular in China, but was also introduced to Korea and Japan. Later, the horizontal *konghou* was gradually replaced by *guqin* and *guzheng*, and was lost to the world in the Song Dynasty.

The vertical *konghou*

The vertical *konghou* (also called Hu-*konghou*) is a stringed instrument with a history of more than 5,000 years, and it was introduced from Central Asia and India through Persia. The vertical *konghou* first appeared in the Eastern Han Dynasty (25 AD-220 AD). The vertical *konghou* was long and curved with 22 strings, and was vertically held in the arms and plucked with two hands. The vertical *konghou* was popular from the Jin Dynasty (266 AD-420 AD) to the Tang Dyansty, and was introduced to Japan in the Tang Dyansty.

The phoenix-headed *konghou*[2]

The phoenix-headed *konghou* was introduced from India in the Eastern Jin Dyansty (317 AD-420 AD), and it had a phoenix or dragon-head on the top of the instrument for ornamentation. The shape of the phoenix-headed *konghou* is similar to that of the vertical *konghou*.

The modern *konghou*

Konghou has been revived in the 20th century as a double bridge. In 1978, Han Qihua, the instrument maker, from the Shenyang Music Instrument Factory, made the first successful *konghou* with double-row strings. In 1984, Zhang Kun, from the Shenyang Music Conservatory, finally made a new kind of yanzhu modulated *konghou* (雁柱转调箜篌) which brought new life to the ancient instrument. This new type of *konghou* is two meters high, 90 meters long and 20 meters wide, and features 88 strings.

The modern *konghou* and Western harp

Konghou was introduced from Persia, and to the east to China, it was called *konghou* and from the west to Europe, called harp. The main feature that distinguishes the modern *konghou* from the Western harp is that the harp is a single-sided case (琴箱) with single-row strings, while the modern *konghou* has double-sided cases with double-rowed strings, and has 36 strings on each side. The playing methods of *konghou* are similar to the harp's. When playing *konghou*, the thumb, index finger, middle finger, ring finger of left and right hands are responsible for playing their respective (各自的) strings. Since there are two rows of strings, it is similar to having two harps and the modern *konghou* can be used in solos, ensemble or accompaniment.

III. Performing Techniques

The basic playing postures of the modern *konghou* are divided into sitting posture and standing posture, and the former is the main one. The playing methods of *konghou* are similar to the harp's, and the main performing techniques are:

琶音 arpeggio

颤音 vibrato

上滑音 up-portamento

下滑音 down-portamento

按音 press

掐音 pinch

散板 free rhythm

刮奏 glissando

扫弦 sweep

止音 muffling

泛音 harmonic/overtone
回滑音 turn
接近琴码演奏 sul ponticello

During playing, besides the fingertips of the left and right hands, the right hand can play with plectrum (or pick).

IV. Musical Classics

Spring on a Moonlit River	《春江花月夜》
High Mountains and Flowing Waters	《高山流水》
The Fisherman's Evening Song	《渔舟唱晚》
River Scene on the Qing Ming Festival	《清明上河图》
Longing for the Secular Life	《思凡》
The Floating Clouds	《云》
Silver Clouds chasing the Moon	《彩云追月》
The Vivid Reflection	《清影》
Autumn Moon over the Calm Lake	《平湖秋月》
The Dance Music of the Yi People	《彝族舞曲》

Terms

sound-box	/saʊnd bɑːks/	共鸣箱
column/ pillar	/ˈkɑːləm/, /ˈpɪlər/	立柱
scroll	/skrəʊl/	琴头
S-shaped panel/ board	/es ʃeɪpt ˈpænl/, /es ʃeɪpt bɔːrd/	S形面板
handle of modulation	/ˈhændl əv ˌmɑːdʒəˈleɪʃn/	转调手柄
tuning pin	/ˈtuːnɪŋ pɪn/	琴钉
bridge	/brɪdʒ/	弦马；琴马
string	/strɪŋ/	琴弦
bridge pin	/brɪdʒ pɪn/	弦钉
arpeggio	/ɑːrˈpedʒioʊ/	琶音
vibrato	/vɪˈbrɑːtəʊ/	颤音
up portamento	/ʌp ˌpoʊrtəˈmentoʊ/	上滑音
down portamento	/daʊn ˌpoʊrtəˈmentoʊ/	下滑音
press	/pres/	按音
pinch	/pɪntʃ/	掐音
free rhythm	/friː ˈrɪðəm/	散板

glissando	/glɪˈsændəʊ/	刮奏
sweep	/swiːp/	扫弦
muffling	/ˈmʌflɪŋ/	止音
harmonics/ overtone	/hɑrˈmɑnɪks/, /ˈoʊvərtoʊn/	泛音
turn	/tɜːrn/	回滑音
sul ponticello	/ˈsʌl pɒntɪˈtʃeloʊ/	接近琴马演奏

Notes on the text

1. phoenix-headed *konghou* 凤首箜篌
2. *Peacocks Fly from the East to the South*...and could recite poems at 16 《孔雀东南飞》中曾叙"十三能织素,十四学裁衣,十五弹箜篌,十六诵诗书。"

Exercises

I. Comprehension questions

1. How many shapes of ancient *konghou* are there?
2. When was *konghou* playing most prevalent?
3. When was *konghou* "revived"?
4. What is the main feature that distinguishes the contemporary *konghou* from the Western harp?
5. How is the modern *konghou* played?

II. Translating useful expressions

1. 箜篌是中国古代的弹拨乐器,在明朝后失传。
2. 箜篌的演奏技巧和竖琴相似。
3. 现代箜篌是两排弦,每两条对应的琴弦发音相同。
4. The modern version of *konghou* does not resemble the ancient one, but its shape is similar to Western concert harps.
5. The main playing techniques of *konghou* have trill, slide, vibrato, arpeggio, roll and harmonic.
6. The two rows of strings start from a tuning peg and beyond the playing area

travel over two bridges on opposite sides of the instrument, and are then fixed at the far end to opposite sides of a freely moving lever so that depressing one of the string pairs raises the pitch of the other.

III. Brainstorm

Konghou playing was most prevalent in the Sui and Tang Dynasties, but *konghou* became extinct sometime in the Ming Dynasty. Not until the 20th century was it revived, but the modern version of *konghou* does not resemble the ancient one. **Innovation is the primary impetus of the development**. From the development of *konghou*, give a specific example to discuss the importance of innovation to everything.

Lesson Three
Bowed-String Instrument

Lesson Three
Bowed-String Instrument

Erhu

Erhu is the chief bowed-string instrument in the Chinese orchestra. Characterized by its versatile playing technique, *erhu*, which is often associated with sorrow, can produce the most heart-wrenching sounds. It is sometimes known as the Chinese violin or Chinese two-stringed fiddle.

The instrument is termed *erhu*, for it has two strings— "*er*" meaning two, "*hu*" indicating its foreign origin. In ancient China, the Hu people commonly referred to barbarians. The name *huqin* (胡琴) literally means "instrument of the Hu people", suggesting that the instrument may have originated from regions to the north or west of China generally inhabited by nomadic people on the extremities of past Chinese kingdoms.

It is used as a solo instrument as well as in small ensembles and large orchestras. As a very versatile instrument, *erhu* is used in both traditional and contemporary music arrangements, such as in pop, rock and jazz. The usual playing range of the instrument is about two and a half octaves. The maximum range of it is three and a half octaves.

I. Construction

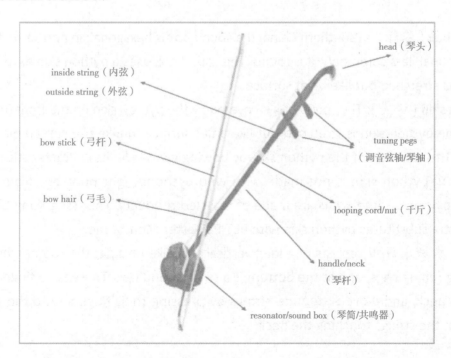

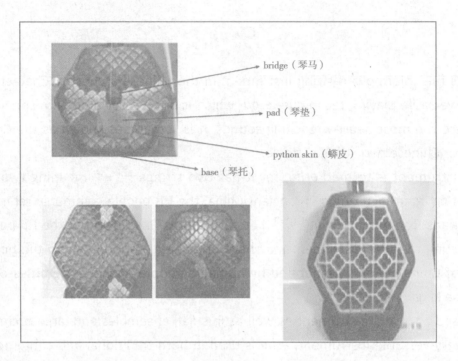

This usually homophonic instrument is played with a bow which is trapped in between the instrument's two strings. The bow is usually made of horsehair. The rosin-lathered horsehairs' movement against the strings produces soul-stirring sounds through left-right bowing actions. The absence of a fingerboard renders the instrument's pitch more difficult to control when bowing, but at the same time allows the instrument to have greater gradations in tone and a richer palette of tone colors.

Sound box (琴筒): In southern China, the sound box is hexagonal; in northern China, it is octagonal; less commonly, it is round. The sound box uses a python skin as its "top", which is stretched out like a flat surface.

Python skin (蟒皮): The sound box is covered by the python skin on the front (playing) end. The python skin is stretched out like a flat surface, unlike the curved plate of a violin. The vibration of the python skin by bowing gives *erhu* its characteristic sound. The *erhu* python skin is not made with wild pythons right now, but from farm-raised pythons. Some *erhus* are made of recycled products. The Hong Kong Chinese Orchestra substitutes python skin with PET Polyester Membrane.

Neck (琴杆): *Erhu* consists of a long vertical stick-like neck, at the top of which are two big tuning pegs, and at the bottom is a small sound box. There is no fingerboard on the neck, and *players stop* the strings by pressing their fingers onto the strings without the strings touching the neck.

Head (琴头): It usually has a simple curve with a piece of bone or plastic on top, but is sometimes elaborately carved with a dragon's head.

Tuning pegs (琴轴): They are at the top of the head, traditional wooden, or metal machine gear pegs. Two strings are attached from the pegs to the base.

Nut/looping cord (千斤): It is made from string, or, less commonly, a metal hook. It is placed around the neck and strings, acting as a nut. It pulls the strings towards the skin, holding a minute wooden bridge in place.

Strings (琴弦): There are two strings, the inner string and the outer string. They are very close to each other although the player's left hand in effect plays as if on one string. The two strings of *erhu* are usually tuned a fifth apart, with the inner string tuned to the lower pitch and the outer string tuned to the higher one. The strings were formerly made of twisted silk but today are usually made of metal. The use of steel strings replaced the use of silk strings gradually. By 1950, the thinner outer string had been replaced by a violin E-string with the thicker inner string remaining silk. By 1958, professional players used two steel strings to replace both two strings as standard.

Bridge (琴马): The bridge of *erhu* is made from wood. It has a flat base and doesn't require "fitting" onto the instrument, because *erhu* uses python skin as its "top".

Bow (琴弓): *Erhu*'s bow has screw device to vary bow hair tension. The horsehair bow is never separated from the strings, instead, it passes between them as opposed to pass over them as Western bowed stringed instruments. Both sides of bow hair are used while bowing. The hairs are slackened. The bow hand is used to press the hairs away from the bow stick to create enough tension to stroke the strings properly.

The bow stick is made from bamboo, while the bow hair is usually white horsehair.

Pad (琴垫): The pad is a piece of sponge, felt, or cloth placed between the strings and skin below the bridge to improve its sound.

Base (琴托): It is a piece of wood attached to the bottom of the sound box to provide a smooth surface to rest on the leg.

II. Development

According to records, *huqin* first appeared in the Tang Dynasty as an instrument called *xiqin* (奚琴); it used a bamboo strip in the middle of its two strings. *Xiqin* is believed to have originated from the Xi people located in current northeast of China.

During the Song Dynasty, *huqin* with a bow made from horsetail hair was already in use around the northwestern regions. During the Yuan Dynasty, the Mongolian tribes used *Hu Weng Er* in religious festivals and in the military.

In the Ming Dynasty, *erhu* became popular nationwide, and the main instrument in folk accompaniment and ensemble. The way of trapping horsehair in between the instrument's two strings was created.

Over the course of a thousand years from when it had been first played, *huqin* has evolved and developed into numerous other variations like the pimo *erhu* (皮膜二胡), *jinghu* (京胡), *jing erhu* (京二胡), *ruangong erhu* (软弓二胡), *yuehu* (粤胡), *sihu* (四胡), *banmian banhu* (板面板胡) and *yehu* (椰胡).

Erhu was standardized until the early 1900s. The development of *erhu* today is largely credited to Liu Tianhua[1]; it was with Liu that *erhu*, previously an exclusive ensemble instrument, gained a stronger repertoire and playing technique. With Liu's new developments, *erhu* quickly became the most outstanding and representative of all the bowed stringed instruments.

III. Performing Techniques

Erhu is played sitting down. The sound box is placed on the top of the left thigh. The neck is held vertically.

Right-hand techniques / bowing

Bowing techniques are sometimes called right-hand techniques as the bow is gripped at its right corner by the right hand. The thumb is placed on the bamboo stick of the bow, and the middle and ring fingers are positioned on the bow hairs. The fingers pull at the bow with the fluid motion of the wrist and forearm to play. Below are some of the more common bowing techniques used by *erhu*:

down bowing and up bowing (推弓、拉弓): The bow hairs, trapped between the strings, are bowed along the qintong with smooth left and right movements.

détaché (分弓): The use of one bowing movement for a single note. Stronger notes are affected using the full extent of the bow in one clean sweep.

fast bowing (快弓): A fast version of separate bowing. It ensures a rapid playing of notes to produce detached sounds.

tremolo (抖弓): Tremolos are produced using the wrists and arms in trembling motions to play the same note at fast speeds using the tip of the bow.

slurred / legato (连弓): The playing of two or more notes using a single pull of the bow.

spiccato (跳弓): The technique of the bow bouncing on the strings, with one note per bounce.

staccato (顿弓): A short and detached note played at one note per bow. Unlike spiccato, this technique does not have bounce.

Left-hand techniques / fingering

Erhu uses the index, middle, ring and last fingers of the player's left hand to play. Some left-hand techniques are detailed below:

position (把位): The arrangement of different pitches on the strings.

vibrato (揉弦): Vibrato can be practiced by two main ways: move wrist up and down while pressing at the strings or use finger pressure to suppress the string, increasing and decreasing its tension.

portamento (滑音): There are four types of glides: upward portamento, downward portamento, backward portamento (回滑音), portamento with fingers in close postion (垫指滑音).

harmonics (泛音): Harmonics on *erhu* are akin to those of the violin. It also includes natural harmonics and artificial harmonics.

pizzicato (拨弦): Using the left hand to pluck the strings. This technique is usually performed on *erhu*'s free strings, as no pressing of strings is required.

IV. Musical Classics

Moon Night	《月夜》
Flickering Red of Candle	《烛影摇红》
Serenity	《病中吟》
March Toward Brightness	《光明行》
Moon Reflected on Erquan Spring	《二泉映月》
Horse Race	《赛马》
Spring on a Moonlit River	《春江花月夜》
Soughing Pines	《听松》
Autumn Moon over the Calm Lake	《平湖秋月》
Flowing Rivers	《江河水》
Henan Folk Tune	《河南小曲》

Little Flower-Drum 《小花鼓》
On the Grassland 《草原上》
Sanmen Gorge Rhapsody 《三门峡畅想曲》

Notes on the text

Liu Tianhua 刘天华（1895—1932），中国近代作曲家、演奏家、音乐教育家。

Terms

tremolo	/ˈtremələʊ/	抖弓
staccato	/stəˈkɑːtəʊ/	顿弓
fast bow	/fæst baʊ/	快弓
pizzicato	/ˌpɪtsɪˈkɑːtəʊ/	拨弦
harmonics	/hɑrˈmɑnɪks/	泛音
détaché	/detaʃe/	分弓
legato	/lɪˈɡɑːtəʊ/	连弓
vibrato	/vɪˈbrɑːtəʊ/	揉弦
portamento	/ˌpoʊrtəˈmentoʊ/	滑音
trill	/trɪl/	颤音
acciaccatura	/ɑˌtʃakəˈturə/	打音
spiccatto	/spɪˈkɑːtəʊ/	跳弓
sound box/ resonator body	/saʊnd bɑːks/, /ˈrezəneɪtər ˈbɑːdi/	琴筒
python skin	/ˈpaɪθɑːn skɪn/	琴皮；蛇皮
neck/ handle	/nek/, /ˈhændl/	琴杆
head/ top/ tip of neck	/hed əv nek/, / tɑːp əv nek/, / tɪp əv nek/	琴头
tuning peg	/ˈtuːnɪŋ peɡ/	琴轴
nut/ looping cord	/nʌt kɔːrd/, /ˈluːpɪŋ kɔːrd/	千斤
inner string	/ˈɪnər strɪŋ/	内弦
outer string	/ˈaʊtər strɪŋ/	外弦
bridge	/brɪdʒ/	琴马

Exercises

I. Comprehension questions

1. How did *erhu* get its name?
2. When and where did *erhu* first appear?
3. What music can *erhu* play?
4. How does *erhu* make sound?
5. What is the correct posture to play *erhu*?

II. Translating useful expressions

1. 二胡是中国民族乐团中主要的弓弦乐器。
2. 琴弓的弓毛夹置于两弦之间。
3. 二胡通过推拉弓擦弦后振动琴皮发音。
4. Characterized by its versatile playing technique, *erhu*, which is often associated with sorrow, can produce the most heart-wrenching sounds.
5. With Liu Tianhua's new developments, *erhu* quickly became the most outstanding and representative of all the bowed stringed instruments.
6. The absence of a fingerboard renders the instrument's pitch more difficult to control when bowing, but at the same time allows the instrument to have greater gradations in tone and a richer palette of tone colors.

III. Brainstorm

Please do research and compare *erhu* with Western fiddles comprehensively and report what you have found to your classmates.

Lesson Four
Hammered-String Instrument

Lesson Four
Hammered-String Instrument

Yangqin

Yangqin, a Chinese hammered dulcimer, derived from Iranian santur, is a percussion-stringed instrument with a wooden trapezoidal resonant soundboard. It used to be written with the characters 洋琴 (lit. "foreign zither"), but in 1910, the first character changed to 扬 (also pronounced "*yáng*"), which means "acclaimed". *Yangqin* has the characteristics of percussion and string instrument, and has been called the "Chinese piano" as it has an indispensable role in the accompaniment of Chinese string and wind instruments.

I. Construction

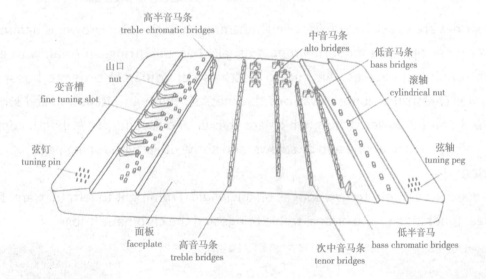

扬琴正面图

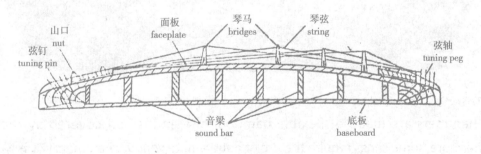

扬琴剖面图

As a type of hammered instrument, *Yangqin* shares many elements of construction of the hammer-stringed family, including soundbox, nut, tuning screw, fine tuner, bridge, string, hammer and cylindrical nuts and so on.

String

The strings are struck by two lightweight bamboo beaters (also known as hammers) with rubber tips. The modern *yangqin* usually has 144 strings in total, with each pitch running in courses, with up to 5 strings per course in order to boost the volume. The strings come in various thicknesses: the high-pitched naked strings (裸弦) and the lower strings, which are thicker and wound (缠弦) with copper. The strings are tied at one end by screws, and at the other with tuning pegs.

Bridge

There are usually four to five bridges on a *yangqin*. From right to left, they are: bass bridge, tenor bridge, alto bridge, treble bridge and the chromatic bridge.

Hammer

The hammers are made of flexible bamboo, one end is half covered by rubber. Furthermore, some songs require the use of double-note *yangqin* hammers (双音琴竹), which have 2 striking surfaces, allowing the player to play up to 4 notes simultaneously.

Lesson Four
Hammered-String Instrument

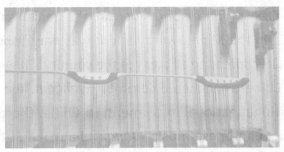

Cylindrical nut

They are fitted into fine-tuning devices. On both sides of a *yangqin*, aside from the tuning screws, are numerous cylindrical metal nuts that can be moved for fine tuning the strings or raise the strings slightly to eliminate unwanted vibrations that may occur.

II. Development

Yangqin is the variant of an international instrument—dulcimer, which has existed since ancient times. Among its forms are hammered delcimers and fretted dulcimers. The hammered ones are free-standing, frequently trapezoidal in shape, with many strings struck by handheld "hammers". It is called santur in the Middle East, cimbalom or tympanon[1] in Europe. The fretted ones usually have three or four strings, generally played on the lap by strumming; including Appalachian dulcimer[2], Banjo dulcimer[3], Resonator dulcimer, Bowed dulcimer and Electric dulcimer.

There are mainly three theories to explain how the instrument was introduced to China: a. introduced by land, through the Silk Road; b. introduced by sea, through the port of Guangzhou (Canton); c. invented without foreign influence by the Chinese themselves.

As a likely predecessor of *yangqin*, the Iranian santur, was introduced to China from Persia in the Ming Dynasty and it was only popular in Guangzhou(Canton) at first, and later spread to all parts of the country. Chinese musicians played it and rebuilt it as a Chinese hammered dulcimer—*yangqin*, and since then it has become a Chinese traditional national instrument. In China, there are four main genres of *yangqin* music: Cantonese music *yangqin*, Jiangnan Sizhu *yangqin*, Sichuan *yangqin* and Dongbei *yangqin*.

The traditional *yangqin* has narrow pitch range, small volume and difficult modulation. After many great reforms, the modern *yangqin* has already had five courses of bridges, though traditional instruments, with three or more courses of bridges, are also still widely in use. A pedal-type damper is equipped under the modern *yangqin* to dampen the vibrations and the steel alloy strings replace the traditional bronze strings. All these improvements increase the range of *yangqin* to five octaves and can modulate freely. Today, *yangqin* is the main instrument in Chinese National Orchestra and it is suitable for ensemble, solo, accompaniment and other art forms.

III. Performing Techniques

The *yangqin* performance begins by holding the hammers, one in each hand, to hit the strings alternately. The main performing techniques of *yangqin* include:

Tan (弹): Strike

It is the basic playing technique. By holding the hammers, one in each hand, strikes the strings alternately.

Lun zhu (轮竹): Tremolo

Alternating the hammers in left hand and right hand to strike theh strings rapidly.

Chan zhu (颤竹) : Short tremolo

Flicking the flexible hammers to bounce continuously on strings, resulting in a short, quick tremolo.

Ya rou xian (压揉弦): Vibrato

Trembling the strings on the other side of the course after striking.

Fan zhu (反竹)

Striking the strings with the bamboo side for crisper, more percussive sound, this technique is often utilized in the higher ranges.

Lesson Four
Hammered-String Instrument

Gua zou(刮奏): **Glissando**
Running the ends of the hammers up or down the strings.

Hua yin(滑音): **Portamento**
A glide from one note to another.

Sliding finger tool(滑音指套)
It is worn on a performer's hand during performances execute portamentos and vibratos, producing sharp or penetrating sound, imitating birds twitter etc.

The other important playing techniques include:

单竹 single note
双竹 double notes
拨、勾弦 pizzicato
顿音 staccato
泛音 harmonics
音阶 scale
琶音 arpeggio

IV. Musical Classics

Joyous News	《喜讯》
Three-Six	《弹词三六》
Song of the Border	《边寨之歌》
Picturesque Tianshan Mountain	《天山诗画》
Night Picture of Yaoshan	《瑶山夜画》
A Rhyme of Rivers in Sichuan	《川江韵》
Falling Flowers · Nights	《落花·夜》
A Trip on Ancient Road	《古道行》
Dragon Boat	《龙船》
The General's Command	《将军令》
Lin Chong Flees in the Night	《林冲夜奔》
Ya Lu Zang Bu Riverside	《雅鲁藏布江边》
Romance of the Yellow Earth	《黄土情》
The Phoenix Is Nodding	《凤点头》

Memories	《意事曲》
A Symphonic Poem of Taiwan Strait	《海峡音诗》
Yellow River Capriccio	《黄河随想曲》

Notes on the text

1. Tympanon 斯拉夫、匈牙利、波兰、俄罗斯等国称庭帕农琴，也称 cimbalon 钦巴龙琴。
2. Appalachian dulcimer 阿帕拉契亚扬琴。阿帕拉契亚山脉在北美洲东部，南起美国阿拉巴马州，北至加拿大拉布拉多省。
3. Banjo dulcimer 班卓琴

Terms

soundbox	/ˈsaʊndbɑːks/	共鸣箱
nut	/nʌt/	山口
tuning screw	/ˈtuːnɪŋ skruː/	弦钉
fine tuner	/faɪn ˈtuːnər/	滚轴
hammer	/ˈhæmər/	琴竹
string	/strɪŋ/	弦
cylindrical nut	/səˈlɪndrɪkl nʌt/	变音槽
panel/ board	/ˈpænl/, /bɔːrd/	面板
bridge	/brɪdʒ/	琴马条
bass bridge	/beɪs brɪdʒ/	低音马条
tenor bridge	/ˈtenər brɪdʒ/	次中音马条
alto bridge	/ˈæltoʊ brɪdʒ/	中音马条
treble bridge	/ˈtrebl brɪdʒ/	高音马条
chromatic bridge	/krəˈmætɪk brɪdʒ/	半音马条
strike	/straɪk/	弹
portamento	/ˌpoʊrtəˈmentoʊ/	滑音
tremolo	/ˈtreməloʊ/	轮
glissando	/glɪˈsændoʊ/	刮奏
vibrato	/vɪˈbrɑːtoʊ/	揉
pizzicato	/ˌpɪtsɪˈkɑːtoʊ/	拨弦
staccato	/stəˈkɑːtoʊ/	顿音
single note	/ˈsɪŋgl noʊt/	单竹
double notes	/ˈdʌbl noʊts/	双竹
staccato	/stəˈkɑːtoʊ/	顿音
harmonics	/harˈmɑnɪks/	泛音
arpeggio	/ɑːrˈpedʒioʊ/	琶音
scale	/skeɪl/	音阶

Lesson Four
Hammered-String Instrument

Exercises

I. Comprehension questions

1. What instrumental category does *yangqin* belong to?
2. How many bridges does a modern *yangqin* have?
3. How many forms do dulcimers have? What are they?
4. What are the four genres of the *yangqin* playing?
5. How many octaves does a modern *yangqin* cover?

II. Translating useful expressions

1. 扬琴是击弦乐器。
2. 扬琴共144根弦，以组布弦，每组弦可多达5根。
3. 扬琴通常有4~5根琴马条，从左至右为：低音马条、次中音马条、中音马条、高音马条、半音马条。
4. The evolution of the psalterium resulted in the harpsichord; that of the dulcimer produced the pianoforte.
5. *Yangqin* is a percussion-stringed instrument with a wooden trapezoidal resonant soundboard.
6. As with a piano, the purpose of using multiple strings per course is to make the instrument louder.

III. Brainstorm

Yangqin is nicknamed "Chinese piano". Please work with your classmates and explain the resemblance of the two instruments, *yangqi* and piano.

Part Two

Western Musical Instruments

Lesson Five

Keyboard Instruments

Lesson Five
Keyboard Instruments

Piano

The piano, belonging to the keyboard instrument, has the reputation of "the king of musical instruments". The piano, invented in Italy by Bartolomeo Cristofori[1] around the year 1709, is shortened from an Italian term ***pianoforte***, and the terms ***piano*** and ***forte*** indicate "soft" and "loud" respectively. Because of its wide pitch range, the piano is widely used in classical, jazz, traditional and popular music for solo, ensemble, accompaniment, and for composing, songwriting and rehearsals. Because of its wide availability, the piano has become one of the Western world's most familiar musical instruments.

I. Construction

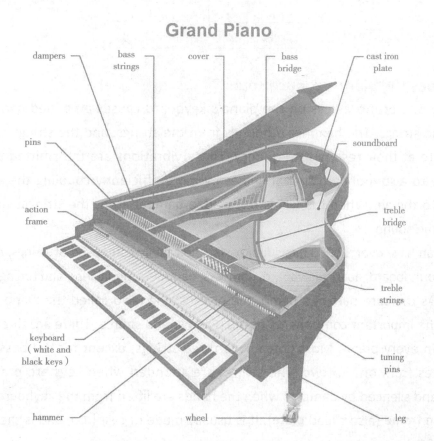

Grand Piano

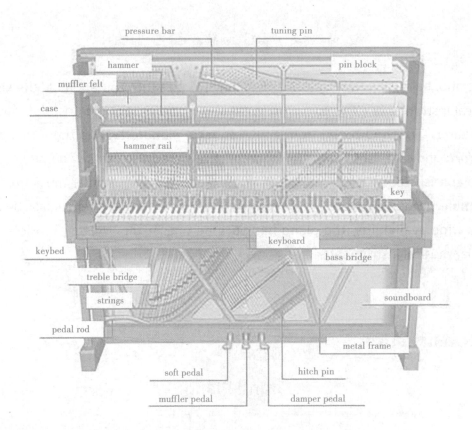

How does the piano produce sound?

Pressing one or more keys on the piano's keyboard causes a padded hammer to strike the strings. The hammer rebounds from the strings, and the strings continue to vibrate at their resonant frequency. These vibrations are transmitted through a bridge to a soundboard that amplifies by more efficiently coupling the acoustic energy to the air. When the key is released, a damper stops the strings' vibration, ending the sound.

Pianos can have over 12,000 individual parts, and mainly composed of strings, cast iron frame, soundboard, action, keyboard, dampers, hammers, pedal, rim, and tuning pins[2].

String: As the core part of a piano, the piano string (also called the piano wire) is one of the important components of the piano sound source. There are three group strings in every piano. Most notes have three strings, except for the bass, which graduates from one to two. The strings are sounded when keys are pressed or struck, and silenced by dampers when the hands are lifted from the keyboard.

Cast iron frame (also called **plate**): It is usually made of cast iron, and is the central frame of the piano, and also the resonance plate of the piano.

Lesson Five
Keyboard Instruments

cast iron plate of a grand piano

Soundboard: The soundboard is a resonator consisting of a thin board whose vibrations reinforce the sound of the instrument. The soundboard posts on the underside of grand piano or back of the uprights piano.

Action: A mechanical device on a piano connects the keyboard to the strings, and it is the soul and the most important part of the piano.

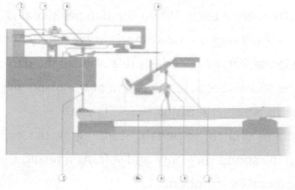

Erard square action

Keyboard: Most modern pianos have a row of 88 black and white keys (52 white keys and 36 shorter black keys which are raised above the white keys, and set further back on the keyboard). Black keys are traditionally made of ebony, and the white keys were covered with the strips of ivory or plastics.

Damper: A damper is a device that deadens, restrains, or depresses the strings. The function of damper is to stop the strings from vibrating and making sound by means of soft and thick tweeds.

Hammer: The hammer of a piano is made of high-quality wool which is condensed into felt and then bonded to wood. For a hammer, its shape, weight and elasticity have a big impact on the volume and timbre, and its function is to strike the strings.

the hammer of a piano

Pedal: Piano pedals are generally made of metal materials, and the piano has three pedals: sustain pedal, soft pedal and middle pedal.

Sustain pedal (damper/right/resonance pedal): It is often simply called "the pedal", since it is the most frequently used. When the damper pedal is pressed down, the damper, which is usually pressed on the string, is immediately raised, which makes all strings continue to vibrate. After the pedal is released, all the dampers are pressed on the strings to stop the sound. Because the sound of the piano will be enlarged a little by pressing this pedal, it is also called loud (强音) pedal.

Soft pedal (left pedal or una corda pedal): Stepping on the soft pedal will reduce the sound and make the sound very pure and soft. As a result, the left pedal is often compared to the "Spinner's microphone"[3].

Middle pedal (a sostenuto/double weak pedal): On grand pianos, the middle pedal is a sostenuto pedal. The sostenuto pedal keeps raising any damper already raised at the moment the pedal is depressed. This makes it possible to sustain selected notes (by depressing the sostenuto pedal before those notes are released) while the player's hands are free to play additional notes (which don't sustain). On many upright pianos, when pressing it, a piece of felt will be dropped between the hammers and strings, greatly muting the sounds.

Rim or case: It is most commonly made of hardwood, typically hard maple or beech, and its massiveness serves as an essentially immobile object from which the flexible soundboard can best vibrate.

II. Development

Origin

The invention of the piano was based on earlier technology in keyboard instruments. The development of pipe organs enabled instrument builders to learn about creating keyboard mechanisms for sounding pitches since ancient time. During the middle ages, there were several attempts at creating stringed keyboard instruments with struck strings, e.g. the hammered dulcimers, the first string instruments with struck strings. By the 17th century, the mechanisms of keyboard instruments such as the clavichord and the harpsichord were well developed. The strings are struck by tangents in a clavichord, while the strings are mechanically plucked by quills when the performer depresses the key in a harpsichord. After centuries of work on the mechanism of the harpsichord in particular, the instrument builders found the most effective ways to construct the case, soundboard, bridge, and mechanical action for a keyboard intended to sound strings.

Invention

About 1709, Bartolomeo Cristofori, who was an expert harpsichord maker and was well-acquainted with the knowledge on stringed keyboard instruments and actions, invented the first piano which had 61 keys and a wooden framework in Italy. Cristofori named the instrument *un cimbalo di cipresso di piano e forte* ("a keyboard of cypress with soft and loud"), abbreviated over time as *pianoforte, fortepiano*, and later, simply, piano.

Cristofori's great contribution to piano was designing a stringed keyboard instrument in which the notes are struck by a hammer. Cristofori's piano action was a model for the many approaches to piano actions that followed in the next century.

Cristofori's early instruments were made with thin strings, and had a quieter sound and smaller dynamic range than the modern piano, but they were much louder and with more sustain in comparison to the clavichord—the only previous keyboard instrument. While the clavichord allows expressive control of volume and sustain, it is relatively quiet. The harpsichord produces a sufficiently loud sound, but it does not allow variation in volume. Cristofori's early piano offers the best of both instruments, combining the ability to play loudly and perform sharp accents.

Early piano

Cristofori's new instrument remained relatively unknown, until his diagram of the mechanism was translated into German and then widely distributed. As an organ builder, Gottheson Silbermann's[4] (1683-1753) pianos were totally direct copies of Cristofori's, but with one important addition: Silbermann invented the forerunner of the modern sustain pedal, which lifts all the dampers from the strings simultaneously. This innovation allows the pianist to sustain the notes that they have depressed even after their fingers are no longer pressing down the keys, and player's hands can be relocated to a different register of the keyboard preparing for a subsequent section.

In the late 18th century, piano-making flourished in the Viennese school. Viennese-style pianos were built with wood frames, two strings per note, and leather-covered hammers. Some of these Viennese pianos had the opposite coloring of modern-day pianos; the natural keys were black and the accidental keys white. The pianos of Mozart's day had a softer tone than 21st century pianos, with less sustaining power.

Modern piano

With the help of Industrial Revolution, the Mozart-era piano underwent tremendous changes that led to the modern structure of the instrument in the period from 1790 to 1860. In 1821, the invent of the *double escapement action*[5] incorporated a *repetition lever* (also called the *balancier*) which permitted repeating a note even if the key had not yet risen to its maximum vertical position. The double escapement action gradually became standard in grand pianos, and is still currently incorporated into all grand pianos.

Other improvements of the mechanism included the use of firm felt hammer coverings instead of layered leather or cotton. Felt hammer permits wider dynamic ranges as hammer weights and string tension increased. Invented in 1844, the sostenuto pedal allowed a wider range of effects. One innovation that helped create the powerful sound of the modern piano was the use of a massive, strong cast iron frame. Also called the "plate", the iron frame sits atop the soundboard, and serves as the primary bulwark against the force of string tension that can exceed 20 tons (180 kilonewtons) in a modern grand piano. The increased structural integrity of the iron frame allowed the use of thicker, tenser, and more numerous strings.

A better steel wire was created in 1840. Several important advances included changes to the way the piano was strung. The use of a "choir" of three strings, rather than two for all but the lowest notes, enhanced the richness and complexity of the treble. The implementation of over-stringing[6], in which the strings are placed in two separate planes and each one with its own bridge height, allowed greater length to the bass strings and optimized the transition from unwound (裸) tenor strings to the iron or the copper-wound bass strings. The modern pianos have 88 black and white keys (52 white keys and 36 shorter black keys), so the tonal range of the piano was gradually increased from the five octaves of Mozart's day to the seven octaves (or more) range found on today's pianos. The modern piano comes into being in romantic period, and it almost hasn't had any changes since then.

In the 19th century, as the extension of the capacities of the piano, many virtuoso pianists rose and the piano became a very popular middle-class family's instruments. Music publishers produced many musical works for piano, so piano music was the only medium of domestic music to outsell vocal music in romantic period. Since then, the piano has been the most important family keyboard instrument in Europe and the United Stated.

III. Types

There are mainly two types of pianos: grand (triangular) and upright (vertical), with various styles of each. There are also specialized and novelty pianos, electric pianos, and digital pianos.

In grand pianos the frame and strings are horizontal, with the strings extending away from the keyboard. The action lies beneath the strings. There are multiple sizes of grand piano: baby grand, parlor grand, boudoir grand and concert grand. Grand pianos are used for classic solos, chamber music and art songs, usually in Jazz and pop concerts.

Upright pianos, also called vertical pianos, are more compact due to the vertical structure of the frame and strings. The upright piano is the most popular because of its small size and its price, which makes it suitable for domestic music-making and practice in private homes. Upright pianos are widely used in churches, community centers, schools, music conservatories and practice instruments.

IV. Performing Techniques

Scale
The 24 major and minor scales are very important. When playing the scale, it is important to make all sounds sound the same. The tempo and dynamics must be distributed evenly, and you must pay attention to the control power of your fingers, otherwise, it is difficult to play a satisfactory tone.

Arpeggio
It includes 24 major and minor arpeggios, dominant 7th arpeggios[7] and diminished 7th arpeggios[8]. Arpeggio is based on the scale. It is helpful to improve the ability of expanding and retracting your fingers if you can practice arpeggio well. The arpeggio in piano occupies a large part, so it is necessary and useful to play them well.

Octave
When playing octave, it is very important for your wrist to be very flexible. Octave has two playing techniques: wrist-only; and with the arm and wrist together.

Chord
No matter how many notes of a chord are depressed, they must be simultaneous and neat. The chords of *piano* (P) should be very light and clean, and the chords of *forte* (F) must have a metallic sound.

The third and sixth chord
As one of the very difficult techniques, it refers to the flexibility of the fingers. In playing the technique, the right hand is a little better, but the left hand is very difficult.

Other important techniques
Tremolo, legato, big leap, trill, repeated notes, glissando and high-difficulty finger independence practice[9], etc.

Although technique is often viewed as only the physical execution of a musical idea, many performers stress the interrelationship of the physical and mental (or emotional) aspects of piano playing. You should have some basic and correct knowledge about your deportment, gestures, wrist, hand, fingers and fingering, and you must have some knowledge of music theory and music history. Although playing techniques are very important, being an excellent pianist, you must combine your emotion with piano playing.

Lesson Five
Keyboard Instruments

V. Musical Classics

Classic solos 经典独奏曲

Invention Johann Sebastian Bach （《创意曲》约翰·塞巴斯蒂安·巴赫）

The Turkish March Wolfgang Amadeus Mozart
　　　　　（《土耳其进行曲》沃尔夫冈·阿玛多伊斯·莫扎特）

Piano Sonata No. 11 in A Major Wolfgang Amadeus Mozart
　　　　　（《A大调第十一号钢琴奏鸣曲》沃尔夫冈·阿玛多伊斯·莫扎特）

Appassionata Sonata Ludwig van Beethoven
　　　　　（《热情奏鸣曲》路德维希·凡·贝多芬）

Piano Sonata No. 14 in C-Sharp Minor "Moonlight" Ludwig van Beethoven
　　　　　（《升C小调第十四号钢琴奏鸣曲"月光"》路德维希·凡·贝多芬）

Polonaise in A Flat Major - "Heroic" Frederic Frangois Chopin
　　　　　（《降A大调"英雄"波兰舞曲》弗雷德里克·弗朗索瓦·肖邦）

Nocturne in B Flat Minor Frederic Frangois Chopin
　　　　　（《降B小调夜曲》弗雷德里克·弗朗索瓦·肖邦）

Etude in C Minor "Revolution" Frederic Frangois Chopin
　　　　　（《C小调练习曲"革命"》弗雷德里克·弗朗索瓦·肖邦）

Moments of Music No. 3 in F Minor Franz Schubert
　　　　　（《音乐的瞬间第三首，F小调》弗朗茨·舒伯特）

Standchen Franz Schubert 　　　（《小夜曲》弗朗茨·舒伯特）

Dream Song (Scenes from Childhood 7) Robert Schumann
　　　　　（"梦幻曲"《童年情景》之七 罗伯特·舒曼）

Moonlight Achille-Claude Debussy 　（《月光》阿西尔-克劳德·德彪西）

Maid with Flaxen Hair Achille-Claude Debussy
　　　　　（《亚麻色头发的少女》阿西尔-克劳德·德彪西）

Spring Song Jakob Ludwig Felix Mendelssohn Bartholdy
　　　　　（《春之歌》雅克布·路德维希·菲利克斯·门德尔松·巴托尔迪）

Wedding March Jakob Ludwig Felix Mendelssohn Bartholdy
　　　　　（《婚礼进行曲》雅克布·路德维希·菲利克斯·门德尔松·巴托尔迪）

Paganini Etude Three "Bells" Franz Liszt
　　　　　（《帕格尼尼大练习曲之三"钟"》弗朗茨·李斯特）

Hungarian Dance No. 5 in F Sharp Minor Johannes Brahms
　　　　　（《升F小调第五号匈牙利舞曲》约翰内斯·勃拉姆斯）

Prelude No. 2 in C Sharp Minor　　Sergei Vassilievitch Rachmaninoff

　　　　　（《升C小调第二号前奏曲》谢尔盖·瓦西里耶维奇·拉赫玛尼诺夫）

Classic sonatas 经典奏鸣曲

Piano Sonata K330 in C Major　　Wolfgang Amadeus Mozart

　　　　　（《C大调钢琴奏鸣曲K330》沃尔夫冈·阿玛多伊斯·莫扎特）

Sonata for Two Pianos in D Major　　Wolfgang Amadeus Mozart

　　　　　（《D大调双钢琴奏鸣曲》沃尔夫冈·阿玛多伊斯·莫扎特）

Piano Sonata No. 8, Pathetique　　Ludwig van Beethoven

　　　　　（《第八钢琴奏鸣曲悲怆》路德维希·凡·贝多芬）

Moonlight Sonata　　Ludwig van Beethoven

　　　　　（《月光奏鸣曲》路德维希·凡·贝多芬）

Piano Sonata No. 31 in A-Flat Major　　Ludwig van Beethoven

　　　　　（《降A大调第31号钢琴奏鸣曲》路德维希·凡·贝多芬）

Piano Sonata No. 3　　Frederic Frangois Chopin

　　　　　（《第三钢琴奏鸣曲》弗雷德里克·弗朗索瓦·肖邦）

Piano Sonata in B Flat Major　　Franz Schubert

　　　　　（《降B大调钢琴奏鸣曲》弗朗茨·舒伯特）

Piano Sonata No. 2 in B Flat Minor　　Sergei Vassilievitch Rachmaninoff

　　　　　（《降B小调第2钢琴奏鸣曲》谢尔盖·瓦西里耶维奇·拉赫玛尼诺夫）

Sonata in G Minor　　Domenico Scarlatti　　（《G小调奏鸣曲》多梅尼科·斯卡拉蒂）

Sonata No. 7　　Sergei Sergeyevich Rokofiev

　　　　　（《第七奏鸣曲》谢尔盖·谢尔盖耶维奇·普罗科菲耶夫）

Classic concertos 经典协奏曲

Piano Concerto No.20 in D Minor K.466　　Wolfgang Amadeus Mozart

　　　　　（《D小调第二十号钢琴协奏曲》沃尔夫冈·阿玛多伊斯·莫扎特）

Piano Concerto No.5 in E Flat Major op.73 ||Emperor||　　Ludwig van Beethoven

　　　　　（《降E大调第五号钢琴协奏曲「皇帝」》路德维希·凡·贝多芬）

Concerto for Piano and Orchestra No. 1 in E Minor Op.11　　Frederic Frangois Chopin

　　　　　（《E小调第一钢琴协奏曲》弗雷德里克·弗朗索瓦·肖邦）

Piano Concerto No. 2 in F Minor　　Frederic Frangois Chopin

　　　　　（《F小调第二钢琴协奏曲》弗雷德里克·弗朗索瓦·肖邦）

Piano Concerto No.1 in E Flat Major　　Ferenc Liszt

　　　　　（《降E大调第一钢琴协奏曲》弗朗茨·李斯特）

Lesson Five
Keyboard Instruments

Piano Concerto in A Minor op.54   Robert Schumann
(《A 小调钢琴协奏曲》罗伯特·舒曼)

Piano Concerto in A Minor op.16   Edvard Grieg
(《A 小调钢琴协奏曲》爱德华·格里格)

Piano Concerto No.1 in B Flat Major op.23   Peter Ilyich Tchaikovsky
(《降 B 大调第一号钢琴协奏曲》彼得·伊里奇·柴可夫斯基)

Piano Concerto No.3 in D Minor op.30   Sergei Vassilievitch Rachmaninoff
(《D 小调第三号钢琴协奏曲》谢尔盖·瓦西里耶维奇·拉赫曼尼诺夫)

Piano Concerto No. 2 in C Minor   Sergei Vassilievitch Rachmaninoff
(《C 小调第二钢琴协奏曲》谢尔盖·瓦西里耶维奇·拉赫玛尼诺夫)

Piano Concerto No. 1 in D Minor   Johannes Brahms
(《D 小调第一钢琴协奏曲》约翰内斯·勃拉姆斯)

Piano Concerto in F Major   George Gershwin
(《F 大调钢琴协奏曲》乔治·格什温)

Piano Concerto No. 3   Béla Viktor János Bartók
(《第三钢琴协奏曲》贝拉·维克托·亚诺什·巴托克)

Terms

Dynamics 力度		
pianissimo (pp)	/ˌpiːəˈnɪsɪməʊ/	很弱
piano (p)	/piˈænəʊ/	弱
mezzo piano (mp)	/ˌmetsəʊ piˈænəʊ/	中弱
diminuendo (dim)	/dɪˌmɪnjuˈendəʊ/	渐弱
mezzo forte (mf)	/ˌmetsəʊ fɔːrt/	中强
forte (f)	/fɔːrt/	强
fortissimo (ff)	/fɔːrˈtɪsɪməʊ/	很强
forte (Più f)	/fɔːrt/	更强
crescendo (cresc)	/krəˈʃendəʊ/	渐强

Tempo 速度		
grave	/greɪv/	慢速
largo	/ˈlɑːrgəʊ/	庄严的慢板
lento	/ˈlentəʊ/	广板；广阔地

adagio	/əˈdɑːdʒiəʊ/	从容的；慢板
andante	/ænˈdænteɪ/	行板
allegro	/əˈlegrəʊ/	快板
accel (abbr. of accelerado)	/əkˈsɛl/	加快的
vivace	/vɪˈvɑːtʃeɪ/	较活泼的速度；快板
prestissimo	/prɛsˈtɪsə,mo/	急板

Construction 组件

natural key (white key)	/ˈnætʃrəl kiː/	全音键
accidental key (black key)	/ˌæksɪˈdentl kiː/	半音键
cast iron frame (plate)	/kæst ˈaɪərn freɪm/	铸铁支架
string	/strɪŋ/	琴弦
soundboard	/ˈsaʊndbɔd/	音板
action	/ˈækʃn/	击弦机
keyboard	/ˈkiːbɔːrd/	键盘
damper	/ˈdæmpər/	制音器
hammer	/ˈhæmər/	琴槌
pedal	/ˈpedl/	踏板
sustain pedal (damper/ right/ resonance pedal)	/səˈsteɪn ˈpedl/	延音踏板；右踏板；共鸣踏板
soft pedal (left pedal/ una corda pedal)	/sɔːft ˈpedl/	柔音踏板
middle pedal (a sostenuto/ double weak pedal)	/ˈmɪdl ˈpedl/	延音踏板；消音踏板；倍弱音踏板
rim (case)	/rɪm/	琴箱

Playing techniques 演奏技法

scale	/skeɪl/	音阶
arpeggio	/ɑːrˈpedʒiəʊ/	琶音
octave	/ˈɑːktɪv/	八度
chord	/kɔːrd/	和弦
tremolo	/ˈtremələʊ/	轮指
legato	/lɪˈɡɑːtəʊ/	连音
staccato	/stəˈkɑːtəʊ/	断奏；断音

Lesson Five
Keyboard Instruments

run	/rʌn/	急奏
touch	/tʌtʃ/	触键
glissando	/glɪˈsændəʊ/	刮奏
big leap	/bɪg liːp/	大跳
rubato	/ruˈbɑtoʊ/	自由速度
alberti bass/ broken chord	/alˈbɛrti beɪs/, /ˈbrəʊkən kɔːrd/	分解和弦
let the weight of hand and arm rest on the key		手与臂的力量落在键上
off the key	/ɔːf ði kiː/	手指离键
arm-swing exercise	/ɑːrm swɪŋ ˈeksərsaɪz/	吊臂练习
arm movement	/ɑːrm ˈmuːvmənt/	手臂动作
hand motion	/hænd ˈməʊʃn/	手的运动
forearm rotation	/ˈfɔːrɑːrm rəʊˈteɪʃn/	小臂旋转
crossing hand	/ˈkrɔːsɪŋ hænd/	双手交叉
keep fingers on the key		手指贴键

Notes on the text

1. Bartolomeo Cristofori 巴托罗密欧·克里斯多佛利（1655—1731），意大利人，钢琴发明者，他是意大利佛罗伦萨美第奇家族的一位乐器制作师。
2. tuning pin 调音钉
3. Spinner's microphone 弦乐演奏者的弱音器
4. Gottfried silbermann 戈特弗里德·西伯尔曼（1683—1753），德国人，18世纪早期的大管风琴制造者。
5. the double escapement action 双擒众击弦机
6. over-stringing: also called cross-stringing 交叉弦
7. dominant 7th arpeggio 属七琶音
8. diminished 7th arpeggio 减七琶音
9. high-difficulty finger independence practice 高难度手指独立练习

Exercises

I. Comprehension questions

1. Who invented the first piano around the year 1709?
2. When did many virtuoso pianists rise and the piano become a very popular middle-class family's instruments?
3. How many keys does a piano have?
4. How many pedals does a piano have? Name them.
5. How is the layout of the strings in modern piano?

II. Translating useful expressions

1. 钢琴属于键盘乐器，音域达七个八度，享有"乐器之王"的美誉。
2. 一台钢琴有 12 000 多个零件，其中主要部件有弦、铸铁支架、音板、击弦机、键盘、制音器、琴槌、踏板、琴箱和调音钉。
3. 音阶、琶音和三六度和弦都是钢琴很难的演奏技巧。
4. The strings are struck by tangents in a clavichord, while they are mechanically plucked by quills when the performer depresses the key in a harpsichord.
5. Pressing one or more keys on the piano's keyboard causes a padded hammer to strike the strings. When the key is released, a damper stops the strings' vibration, ending the sound.
6. When the sustain pedal is pressed down, the damper, which is usually pressed on the string, is immediately raised, which makes all strings continue to vibrate. After the pedal is released, all the dampers are pressed on the strings to stop the sound.

III. Brainstorm

The piano combines the features of string instruments, percussion instruments and keyboard instruments, so it has wide pitch range and wide availability. From this aspect, brainstorm your understanding to the quotation "***The sea embraces all rivers, and tolerance is the foundation of greatness.***"

Lesson Five
Keyboard Instruments

Accordion

The term "accordion" is derived from the German word "akkord", which means musical chord or concord of sounds. The accordion is a family of the squeezebox, free-reed, aerophone-type instruments. These portable box-shaped instruments involve reeds set into each frame with their own openings. Air from the bellows controlled by the player passes through the openings, which causes the reeds to vibrate and produces musical sounds. It consists of a treble casing with external piano-style keys or buttons and a bass casing usually with buttons attached to opposite sides of a hand operated bellows. The accordion, concertina[1], harmoneon[2] and bandoneon[3] are in the same family. A musician who plays the accordion is called an accordionist.

Although created in Germany in the early 19th century by Christian Friedrich Ludwig Buschmann[4], the accordion may owe its existence to far older free reed instruments such as the Chinese *sheng*[5]. Due to the emigration of Europeans around the world and its association with the common people, it spreads rapidly around the globe and become a staple in music. It could play a huge spectrum of styles from folk, traditional and classical music to jazz and popular music.

I. Construction

How to produce sound?

The accordion consists of the chromatically arranged reeds. This sequence is called the voice. When the bellows is pushed or pulled, the air flows through the specified reeds, which causes vibration of the reeds and produces a specified frequency of the sound. The reeds are mounted on the frame. Each reed's frame has two reeds with the same pitch. When the bellows is pulled, the air flows through the first reed. In opposite way, the air flows through the second reed. Each frame has a valve which is located on the other side of the frame. The valve is responsible for passing the flow of air in one direction in one time.

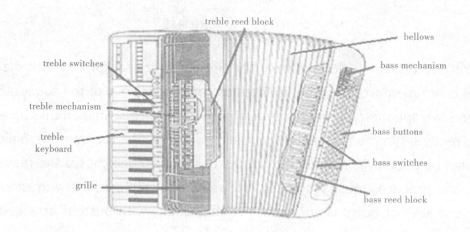

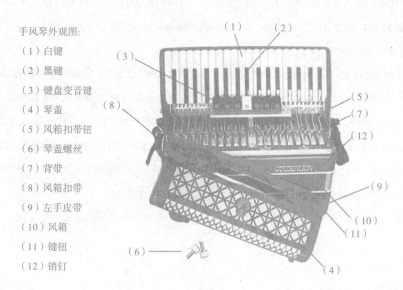

手风琴外观图：
（1）白键
（2）黑键
（3）键盘变音键
（4）琴盖
（5）风箱扣带钮
（6）琴盖螺丝
（7）背带
（8）风箱扣带
（9）左手皮带
（10）风箱
（11）键钮
（12）销钉

Accordions have many configurations and types. Accordions vary not only in their dimensions and weight, but also in number of buttons or keys present in the right-hand and left-hand manuals. Despite these differences, all accordions share some universal components.

Bass: The bass side is located at the left side of instrument. It consists of the bass buttons, which can produce single notes and chords.

Bass buttons: The bass buttons are arranged in rows and columns. The root notes (such as A, C, ...) are generated by buttons located in the first two rows. The chords (A-major, C-minor, ...) are generated by the buttons in the particular column. Bass buttons are arranged in a circle of fifths, this means that, distance between sounds is equal fifth semitones.

Lesson Five
Keyboard Instruments

Below schema presents arrangement of bass buttons in the 120 bass accordion.

Bass buttons, 120 bass accordion

Types of the bass buttons:
1. Major 3rd note
2. Root note
3. Major chord
4. Minor chord
5. Dominant 7th chord
6. Diminished 7th chord

Bellows: Bellows is located at the center of the accordion, between the bass and the melody side and is the "heart" of the accordion. It is made from pleated layers of cloth and cardboard, with added leather and metal. The bellows is used to generate the air pressure which is forced to pass through the internal reeds. The reeds vibrate and produce the sound. The bellows is also the primary means of the articulation of the sound. In this sense, the role of the bellows could be compared to the role of drawing a violin bow against the strings or the role of breathing for a singer.

Reeds: The reed produces the accordions sound. It consists of thin steel, which is riveted at one end of the reed plate, usually made of aluminium. The other end of the reed is free to vibrate in and out of the slot, when air pressure is supplied from the bellows. The reed plates are mounted on reed blocks, which are usually made of poplar wood. Reeds have leathers to moderate the airflow.

reeds

Valves: Valves are the padded bars which open and close sound holes and are operated, through valve levers, from the keys or buttons. These sound holes are located on the treble side of the instrument, under the grille.

Grille: The grille covers the keyboard's treble valves and mechanisms. It is used to decorate the accordion, and usually displays the brand name or the logo of the manufacturer. It is common to see brightly colored grilles with jewels and different colored trimmings. The grille is usually "vented" to allow a louder treble sound. Occasionally, however the grille is used as a muting mechanism.

Switches: The function of a switch is to open or close one or more sets of reeds, via register slides in the reed blocks. The more reeds in an accordion, the more switches become available. Both treble and bass switches are arranged in different orders on different makes of accordions.

Treble keyboard (piano accordion): The treble keyboard has the same layout as the piano. It is located on the right side of the instrument. The keyboard consists of the white keys and black keys arranged chromatically. The lowest note is at the top when you look at it from the front.

Straps: The shoulder straps are used to stabilize an accordion as it is being played. There is one strap for each shoulder. These straps make it possible for a musician to play when standing.

II. Development

Origin

Throughout the Classical and Romantic periods, the humble accordion and its simpler cousin, the concertina, were important parlor, chamber and accompanying instruments. The earliest forms of the accordion were inspired by the Chinese free-reed mouth-blown organ *sheng* which was introduced into Europe in 1777 by a French Jesuit missionary, Pere Amiot[6] from China in the Qing Dynasty. The introduction of *sheng* set off an era of experimentation in free-reed instruments, especially the bellows-operated reed organ such as accordion-type instrument and the mouthblown organ such as harmonica-type instrument. The Germanisches National Museum[7] in Nuremberg holds Europe's largest collection of early German free-reed instruments, accordions, and harmonicas.

Invention

The advent of the accordion is the subject of debate among researchers. The basic form is believed to be first invented by Christian Friedrich Ludwig Buschmann, whose *"handaoline"* was patented in Berlin in 1822. However, there is some dispute about that. Russian researchers claim the earliest simple accordion was first made by Timofrey Vorontsov[8] in 1820.

Some researchers also give credit to Cyril Demian[9], an Austrian-Armenian, who moved to Vienna and worked as an organ and piano maker with his two sons. In 1829, he patented his *"akkordion"*, German for harmony, thus coining its well-known name. Demian's invention is a modification of the *"handaoline"*, which most resembles today's accordion. The instrument comprised a small manual bellows and five keys, although, as Demian noted in his patent description of the instrument, extra keys could be incorporated into the design.

An eight-key bisonoric diatonic accordion (c. 1830s) from A World of Accordions Museum in Wisconsin. Photo by Henry Doktorski.

The instrument only had a left hand buttonboard, with the right hand simply operating the bellows. It was what we now call a push-pull accordion, that produced a different note on each key. His instrument also could sound two different chords with the same key, pulling the bellows a key gives one chord, while pushing the bellows gives the same key a second chord (a bisonoric action). Five keys would give a few notes more than an octave in a diatonic scale and major chords would be easy to produce.

In the 19th century, the accordion eventually supplanted the fiddle as the staple instrument for dance music in northern Europe, because of the relative ease of playing in comparison with the fiddle. Accordion reeds are permanently tuned, so it is hard/impossible to play out of tune, and the arrangement of the keys makes production of major chords very simple.

Early accordions

After Demian's invention, numerous variations of the device soon followed. The English inventor Charles Wheatstone[10] brought both chords and keyboard together in one squeezebox. His *concertina* also had the feature of easily tuning the reeds from the outside with simple tool.

The French Pichenot le Jeune's[11] flutina, which resembles Wheatstone's concertina in internal construction, and has a concertina-like sound, is probably one of the first accordions capable of playing a melody. It is a one-sided bisonoric melody-only instrument with one or two rows of treble buttons, which can play the tonic of the scale as pulling the bellows by left hand. Usually there is no bass keyboard, and the right hand operates the keys. It was a complement of Demian's accordion functionally and when the two instruments combined, a diatonic button accordion was appeared.

concertina

flutina

Innovations

There have been further innovations added to the accordion by now. Various buttonboard and key board systems have been developed too. As technology advances, accordions have also hopped onto the microchip bandwagon. Digital accordions may incorporate electronics to control the tone and volume, as well as the condenser microphones. As a result of this, the accordion can be plugged into a PA system[12] or keyboard amplifier for live shows. The digital accordion is MIDI compatible, which enables the accordion to generate different sound effects.

These days, the accordion is not only used to perform ethnic or folk music, but it is also heard in pop music, rock and pop-rock, and even in sophisticated classical music concerts.

III. Types

Accordions come in many forms and sizes. Depending on the manufacturer and the year of making, accordions may be made from different material, depict different reeds and the number of keys. Because of this, their design and musical sound may vary from one instrument to the next. Some vintage varieties display ornate decorations while newer models depict technology.

There are two main kinds of accordion, distinguished by their different keyboards: button accordions and piano accordions. All accordions have a button keyboard on the left-hand side for bass chords, but on the right-hand side, where the melody is played, they can have either piano or button keys.

A piano-style accordion made it Italy (top) and a Russian Bayan chromatic button accordion (bottom). Photos by Henry Doktorski.

Piano accordion: As the name implies, the piano accordion has a piano-style keyboard. The right-hand treble keyboard has the same layout and design as the one in a regular piano. A full-size accordion has 41 treble keys and approximately 3+ octaves of notes from a low F to a high A. The left-hand side consists of a board of buttons for bass accompaniment. A full-size piano accordion has 120 buttons but there are some varieties that have 140 buttons. The bass system can be free bass, stradella or French 3-3. However, most commonly the bass buttons are configured in the stradella style. The great thing about the piano accordion is that it is very flexible and can be adapted to play any style of music.

Button accordion: A button accordion is a type of accordion on which the treble side or the melody side of the board consists of buttons rather than piano keys. These accordions come in a variety of configurations and styles. However, all button accordions have single notes button on one side and bass and chord buttons on the other side. A further invisible distinction within button accordions is their usage of a chromatic or diatonic buttonboard for the right-hand side.

Diatonic accordion: Most diatonic accordions are button accordions that have one to multiple rows of buttons. The diatonic accordion is smaller than the chromatic. The main difference between a chromatic and a diatonic accordion is that the reeds are bisonoric for diatonic accordions. This means it has two notes per button, depending on whether you push or pull the bellows, thus allowing for it to be smaller. This is the same for bass notes and chords that are different depending on the direction of the bellows. These types of accordions are widely used in folk and ethnic music.

Chromatic accordion: The chromatic accordion is most common in mainland Europe and Russia. Chromatic accordions have buttons for both the right-hand treble side and the left-hand bass side, with the treble side holding a button configuration of 3, 4, or 5 rows of buttons. The treble side usually has a C-system or a B-system. The C-system works very well for playing chords while the B-system is suited for classical music. It is often heard in European folk, pop, and rock.

Unisonoric accordion[13]: Another major difference between accordions is their unisonoric or bisonoric nature. This refers to how the bellows produce notes and pitches by moving the air through the reeds. In a unisonoric accordion, a key or a button produces the same pitch or note regardless of the direction in which the bellows are moving. The pitch of the accordion also depends on the instrument's size.

Bisonoric accordion[14]: A bisonoric accordion, unlike a unisonoric one, produces two different notes or pitches when a button is pressed, depending on the direction of the bellows. When the bellows are pulled out, they make a different note and when they are pushed in, they make a different sound.

IV. Performing Techniques

Holding accordion

Either sitting or standing while holding the accordion, the player should feel at ease to keep balance. Don't slouch over too much, keep the back straight, so that the accordion won't move around too much. Being able to maintain proper balance is crucial. The more evenly balanced the player manages to keep the accordion's

Lesson Five
Keyboard Instruments

weight, the better the player will be able to play because of the added control. And the more control the player has, the less uncomfortable the weight will feel. The instrument should be secured onto the player's chest.

When playing the accordion, straps should be behind the shoulders. The left arm should be between the bellow-strap and the board tighten that strap so that it won't cut off the arm circulation, but not so loose that the hand will slip all over the place.

Playing the accordion

Hold the wrist parallel to the keyboard. The right wrist cannot be bent while keeping the elbow close to the body. The right hand should be free and resting above the piano keyboard. Slip the left hand through the strap that lies below the bass button board and curl the fingers up and over the bass button.

Push down on the air valve. Press the button down softly, and pull the instrument with the left arm. A hissing noise will sound as the air goes into the accordion and the bellows open. No matter how many bass buttons the accordion has, they produce both bass notes and chords. Accordion chord buttons on the left side play three note chords automatically, due to the accordion's internal mechanism. Keep the bass buttons pressed for only a short time, and take the finger off quickly, and a "staccato" is produced.

The basic bellows technique is not running out of air in the middle of a music phrase. To avoid this, emphasis should be placed on changing bellows direction after two short musical phrases. Try lifting the left-hand side of the accordion while squeezing in, so that gravity helps push the bellows together towards the right-hand side. Furthermore, if it is played sitting down, use the right knee higher supporting it on the outward journey and then raise the left knee instead. Expand the instrument's bellows. Softly and evenly push it back together, and press the keys down. Keep pressing the note key while changing directions by pulling the instrument in opposite directions.

Accents: This is when the player suddenly pulls on the bass strap or pushes on the side of the bass board with arm while pressing a note, which creates a short sharp sound. Good coordination is required to get the timing of an accent exact.

Alternating bass: To play this, a root note (fundamental) is played, then a chord of that root note, followed by the fundamental dominant (5th) of the root note, and back to the chord of the root note. A simple example is C, C major, G, G major.

Bellow shakes: The bass arm pulls quickly in and out causing a "stuttering" effect.

There are many types of bellow shakes, which give different effects.

Duple bellow shake: In the space of a beat, a note is sounded twice. The duple bellow shake is played by creating a "hinge" with one side of the bellows and only opening the bellows from the opposite side.

Triple bellow shake: In the space of one beat, a note is sounded three times. The bellow movement is in, out, in and vice versa. This can take a lot of co-ordination to get the feel of the rhythm because the pulse changes bellows direction every time.

Quadruple bellow shake: In the space of one beat, a note is sounded four times. The bellow movement is in, out, in, out. This bellows shake can sometimes be played using the four corners of the bellows, where the bellows are moved in a circular motion, creating the feeling of four counts.

Crescendo (gradually getting louder): This is a term familiar to most musicians, but the accordion is one of the best instruments on which to play a crescendo. Pressure is gradually increased to the bellows, which increase the volume.

Decrescendo (gradually getting softer): This is the opposite of the crescendo, and again the accordion is one of the best instruments on which to perform. By gradually decreasing pressure to the bellows, volume also decreases.

Double action: The pitch of the note is not affected by the direction of the bellows. For example: C is the same note on the out bellows as it is on the in bellows.

Musette: Musette is a sound produced by specifically tuning the reeds of an accordion higher or lower than normal. It is very common in traditional French music. Musette can also be called "tremelo".

Free bass: Unlike the stradella system, all the bass buttons play individual notes. This gives the accordion a fantastic range of notes. Organ and piano pieces can be played without needing to be arranged. Free bass is used by many baroque and classical players.

Stradella bass: This is the traditional bass style of the accordion. There are up to three and some are four rows of buttons which are single notes, and up to four rows of buttons which are fixed chords (3 notes play when one button is played).

V. Musical Classics

Accordion Dance for Accordion and Orchestra   George Antheil

（《手风琴和管弦乐的舞蹈》乔治·安太尔）

Lesson Five
Keyboard Instruments

All Soul's Carnival Henry Brant （《灵魂嘉年华》亨利·布兰特）
American Rhapsody John Serry Sr. （《美国狂想曲》约翰·塞里）
Apocalypse According to St. John Jean Francaix （《圣约翰的启示》让·弗朗赛）
Cantata for the 20th Anniversary of the October Revolution, op. 74 Sergei Prokofiev
（《纪念十月革命胜利20周年的康塔塔》谢尔盖·普罗科菲耶夫）
Four Saints in Three Acts Virgil Thomson
（《四位圣者的三幕剧》维吉尔·汤姆森）
Jazz Suite No. 2 Dmitri Shostakovich
（《爵士第2组曲》德米特里·肖斯塔科维奇）
The Trial of Lucullus Paul Dessau
("Das Verhör des Lukullus"《卢库卢斯的审判》保罗·德绍）
Prelude and Postlude for "Lidoire" Daruis Milhaud
（《"Lidoire"的序幕和终曲》达律斯·米约）

Terms

reed	/rid/	簧片
aerophone	/ˈeroʊˌfoʊn/	气鸣乐器
diatonic	/ˌdaɪəˈtɑːnɪk/	自然音的
chromatic	/krəˈmætɪk/	半音的
bass	/beɪs/	贝司
bass button	/beɪs ˈbʌtn/	贝司键钮
bellows	/ˈbeloʊz/	风箱
valve	/vælv/	气钮
grille	/grɪl/	琴盖；护栅
switch	/ˈswɪtʃ/	键盘变音键
treble keyboard	/ˈtrebl ˈkiːbɔːrd/	键盘
strap	/stræp/	背带
accent	/ækˈsent/	重音
alternating bass	/ˈɔːltərneɪtɪŋ beɪs/	交替低音
bellow shake	/ˈbeloʊ ʃeɪk/	风箱颤音
duple/ triple/ quadruple bellows shakes	/ˈduːpəl ˈbeloʊz ʃeɪks/, /ˈtrɪpl ˈbeloʊz ʃeɪks/, /kwɑːˈdruːpl ˈbeloʊz ʃeɪks/	双重/三重/四重风箱颤音
crescendo	/krəˈʃendəʊ/	渐强
decrescendo	/ˌdiːkrəˈʃendoʊ/	减弱
double action	/ˈdʌbl ˈækʃn/	双重作用
free bass	/friː beɪs/	自由低音
musette	/mjuːˈzet/	缪塞特音
stradella bass	/strəˈdelə beɪs/	斯特德拉低音

Notes on the text

1. concertina 六角形手风琴
2. harmoneon/harmonium 脚踏式风琴
3. bandoneon/bandonion 小六角手风琴
4. Christian Friedrich Ludwig Buschmann 克里斯蒂安·弗里德里希·路德维希·比施曼（德）
5. *Sheng* 笙
6. Pere Amiot 法国传教士阿米奥
7. Germanisches National Museum 日耳曼国家博物馆（德国纽伦堡）
8. Timofrey Vorontsov 蒂莫菲·沃龙佐夫（俄）
9. Cyril Demian 西里尔·德米安（奥地利）
10. Charles Wheatstone 查尔斯·惠斯通（英）
11. Pichenot le Jeune 皮舍诺·热恩（法）
12. PA system 扩音系统
13. unisonoric accordion 拉压同音式手风琴
14. bisonoric accordion 拉压异音式手风琴

Exercises

I. Comprehension questions

1. Which instrument that was introduced from China inspired the invention of accordion?
2. What are the two main kinds of accordion?
3. What is pulled and pushed to cause the air flow inside the accordion?
4. What is the main function of reeds in accordion?
5. How to play the crescendo with an accordion?

II. Translating useful expressions

1. 手风琴是一种带有自由簧片的气鸣乐器。
2. 乐器演奏者拉动风箱产生气流带动簧片震动发出乐音。
3. 19世纪，手风琴最终取代小提琴成为北欧舞曲的主要演奏乐器。
4. The earliest forms of the accordion were inspired by the Chinese free-reed mouth-blown organ *sheng* which was introduced into Europe in 1777 from China in the Qing Dynasty.

5. As technology advances, accordions have also hopped onto the microchip bandwagon.
6. In a unisonoric accordion, a key or button produces the same pitch or note regardless of the direction in which the bellows are moving.

III. Brainstorm

The earliest forms of the accordion were inspired by the Chinese free-reed mouth-blown organ *sheng*. Can you do some research on *sheng* yourself and compare these two instruments? Do you know any other musical instruments or art forms which was also influenced by Chinese musical instruments or art forms?

Electric Organ

The electric organ is an electronic keyboard instrument, also known as electronic organ. It was derived from the harmonium (脚踏风琴), pipe organ and theatre organ. It is named after its resemblances to the organ in terms of performing.

Originally the electric organ was designed to imitate the sounds of organs, or orchestra, but after decades of reformation, now it has timbres of over a thousand kinds, with tone qualities close to acoustic instruments. Furthermore, the tone colors could be edited and stored for individual and personalized needs.

I. Construction

The electric organ has an upright body and three keyboards: the upper keyboard, the lower keyboard and the pedal keyboard. The upper keyboard is usually with 4 or 5 octaves, while the lower keyboard 5 octaves and the pedal keyboard one and a half.

The basic function of the expression pedal is to control the volume, however it is not called the "volume pedal", for the change of volume would lead to changes of emotional expression consequently. Besides, on both sides of the pedal, there are foot switches. The right one is in charge of different timbres, and the left one determines the rhythm, and sometimes the addition of grace notes.

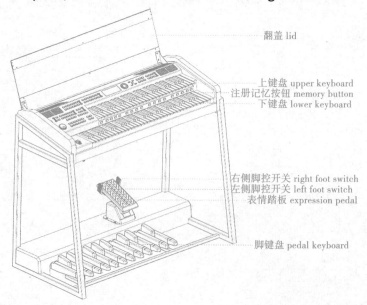

翻盖 lid
上键盘 upper keyboard
注册记忆按钮 memory button
下键盘 lower keyboard
右侧脚控开关 right foot switch
左侧脚控开关 left foot switch
表情踏板 expression pedal
脚键盘 pedal keyboard

Lesson Five
Keyboard Instruments

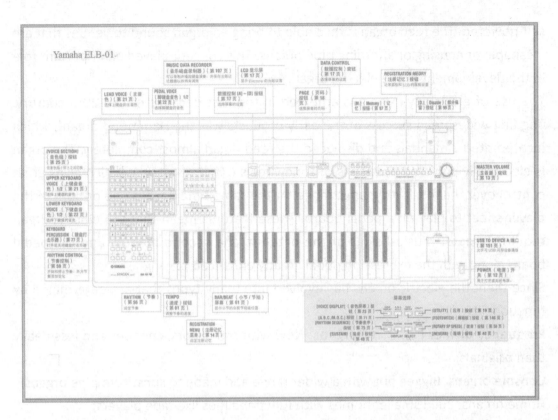

The console surrounding the keyboards consists of many knobs and buttons, which preset, ornament, and edit the music. The buttons are basically in four categories: reverberation module, rhythm module, timbre module, and system module.

II. Development

The predecessors of the electric organ are pipe organs and reed organs.

Until the invention of the reed organ, the pipe organ had been called just the "organ". It is one of the oldest instruments still used in European classical music, and can be drawn back to antiquity. A pipe organ contains one or more sets of pipes, a wind system, and one or more keyboards. The pipes produce sound when pressurized air produced by the win system passes through them. An action connects the keyboards to the pipes. Stops allow the organist to control which ranks of pipes sound at a given time. The organist operates the stops and the keyboards from the console.

The immediate predecessor of the electronic organ was the harmonium, or reed organ, an instrument that was common in homes and small churches in the late 19th and early 20th centuries. Reed organs generate sound by forcing air over a set of reeds by means of a bellows, usually operated by constantly pumping a set of pedals. While reed organs have limited tonal quality, they are small, inexpensive, and self-powered. The reed organ is thus able to bring an organ sound to venues that are incapable of housing or affording pipe organs. This concept played an important role in the development of the electric organ.

The use of electricity in organs emerged in the first decades of the 20th century. The first widespread success of the early products was the Hammond organ[1], which incorporated 2 manuals and displaced the reed organ almost completely. The organ is electrically powered by means of tonewheels (转速脉冲轮), which gives great control over the music's dynamic range. At the same time it frees one or both of the player's feet to play on a pedal board. These features mean that the electric organ requires greater musical skills of the organist, the second manual and the pedal board along with the expression pedal greatly enhanced playing.

Since then, the electric organs has evolved into many types that are determined by functionality:

Frequency divider (分频器) **organs**: Now with transistors, cheaper and less heavy than originals;

Console organs: Bigger, but with a wider range and made to substitute pipe organs;

Home organs: Build smaller homes with functionalities like tape players;

Combo (混合) **/ Spinet** (立式) **/ Transistor** (晶体管): These versions become standardized, probably what most people think when they hear electric organs;

Digital organ: Capable of making all organ sounds digital (using electricity and a speaker);

Synthesizer (合成器) **organ**: Now able to play different instruments using a digital keyboard;

Software organ: Non-physical, you can have any instrument on your computer or phone.

III. Performing Techniques

The performing of the electric organ is compared to "a band played by one", describing not only the effect the instrument could emulate, but also the coordination required of several parts of one's body when playing.

The upper keyboard is usually played by the right hand for the melody, while the lower keyboard by the left hand for accompany, such as harmony, broken chords or simple melodic lines. In specific occasions, both hands could play on the same keyboards, or switch positions on both keyboards.

The pedal keyboard is played by toes and heels. The pedals basically function as another keyboard for the performer. They can be coupled to either of the manuals to play the same register. The pedals mostly are used to play the lowest notes in the piece of music, allowing more complex music to be played on the manuals.

Proper posture is necessary when playing. The focuses are often on the player's upper body, including:

1. Keep straight of the upper body, the hands and arms should be relaxed and in a natural state.
2. Keep the forearm, the wrist and the back of the hand in a horizontal line.
3. Touch the keys with finger pads, except the thumbs.
4. Strike the keys with proper dynamic, neither too heavily nor too lightly.

Other playing techniques are similar to those of other keyboard instruments, such as the piano, including legato, staccato, trill, glissando, chord, etc.

Notes on the text

哈蒙德风琴，以发明者 Lauren Hammond 和 John M. Hanert 命名，通过在电磁拾音器附近旋转金属音轮产生电流来产生声音，然后用放大器加强信号以驱动扬声器箱体。

Terms

lid	/lɪd/	翻盖
upper keyboard	/ˈʌpər ˈkiːbɔːrd/	上键盘
lower keyboard	/ˈləʊər ˈkiːbɔːrd/	下键盘
pedal keyboard	/ˈpedl ˈkiːbɔːrd/	脚键盘
expression pedal	/ɪkˈspreʃn ˈpedl/	表情踏板
foot switch	/fʊt swɪtʃ/	脚控开关
memory button	/ˈmeməri ˈbʌtn/	注册记忆按钮
console	/kənˈsəʊl/	控制台
reverberation module	/rɪˌvɜːrbəˈreɪʃn ˈmɑːdʒuːl/	混响模块
rhythm module	/ˈrɪðəm ˈmɑːdʒuːl/	节奏模块
timbre module	/ˈtæmbər ˈmɑːdʒuːl/	音色模块
system module	/ˈsɪstəm ˈmɑːdʒuːl/	系统模块
legato	/lɪˈɡɑːtəʊ/	连奏
staccato	/stəˈkɑːtəʊ/	断奏
trill	/trɪl/	颤音
glissando	/ɡlɪˈsændəʊ/	刮奏
chord	/kɔːrd/	和弦

Exercises

I. Comprehension questions

1. What was the electric organ derived from?
2. How many kinds of timbre could the electronic organ have?
3. How many keyboards does an electric organ have and what are they?
4. How does a reed organ generate sound?
5. Why is the playing of the electric organ considered "a band played by one"?

II. Translating useful expressions

1. 电子管风琴是电子键盘乐器。
2. 电子管风琴有一千多种音色，其音质接近真实乐器。
3. 电子管风琴被称为"一个人的乐队"。
4. The basic function of the expression pedal is to control the volume, however it is not called the "volume pedal", for the change of volume would lead to changes of emotional expression consequently.
5. At the same time it frees one or both of the player's feet to play on a pedal board.
6. The upper keyboard is usually played by the right hand for the melody, while the lower keyboard by the left hand for accompany, such as harmony, broken chords or simple melodic lines.

III. Brainstorm

Compared with traditional keyboard instruments, the electric organ is more powerful in many ways. Could you come up with some other examples that demonstrate the science changing our way of creating and appreciating arts?

Lesson Six

Stringed Instruments

Violin

The violin, sometimes known as the fiddle[1], is a bowed string instrument and is called "the queen of the musical instruments". The word "violin", meaning "stringed instrument", comes from "Italian violino", first used in English in 1570. The violin typically has four strings, usually tuned in perfect fifths, and is most commonly played by drawing a bow across its strings. As the smallest and highest-pitch instrument in the string family, the range of the violin is over four octaves, and it is often possible to play higher depending on the length of the fingerboard and the skill of the violinist.

Violins are important instruments in a wide variety of musical genres. They are most prominent in the Western classical tradition, both in ensembles (from chamber music to orchestras) and as solo instruments. Violins are also important in many varieties of folk music, including country music, bluegrass music and in jazz.

I. Construction

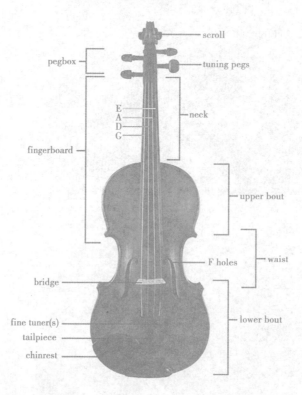

Lesson Six
Stringed Instruments

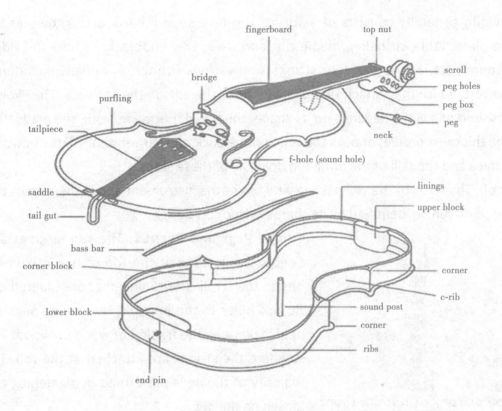

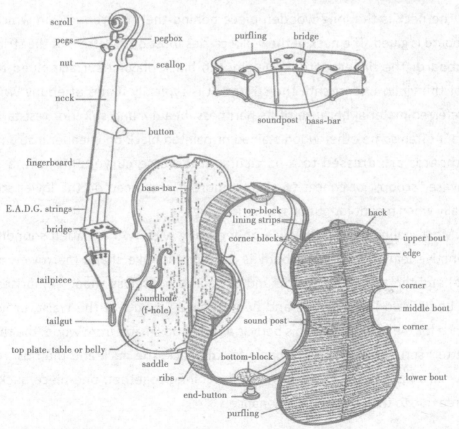

A violin generally consists of a spruce top (the soundboard, also known as the top plate, table, or belly), maple ribs and back, two endblocks, a neck, a bridge, a soundpost, a bass bar, four strings, and various fittings, optionally including a chinrest, which may attach directly over, or to the left of, the tailpiece. The "voice" or sound of a violin depends on its shape, the wood it is made from, the graduation (the thickness profile) of both the top and the back, the varnish that coats its outside surface and the skill of the luthier in doing all of these steps.

Scroll: The scroll of the violin is the very top of the instrument above the pegbox and the scroll can be identified by its characteristic curl design.

scroll and pegbox, correctly strung

Tuning Pegs and Pegbox: The tuning pegs and pegbox are located at the top of the instrument beside the scroll. The tuning pegs are tapered and fit into holes in the pegbox. The tuning pegs are held in place by the friction of wood on wood. This is where the strings are attached at the top. The majority of tuning is performed by tightening and loosening the peg.

Neck: The neck is the long wooden piece behind the fingerboard, on which the fingerboard is glued. The neck of the violin carries most of the stress of the strings.

Fingerboard: The fingerboard is the smooth black playing surface glued to the neck of the violin underneath the strings. It is typically made of ebony which is the preferred material because of its hardness, beauty, and superior resistance to wear, but often some other wood stained or painted black on cheaper instruments. Fingerboards are dressed to a particular transverse curve, and have a small lengthwise "scoop," or concavity, slightly more pronounced on the lower strings, especially when meant for gut or synthetic strings.

Body: Most violins have a hollow wooden body with two f-shaped soundholes. A distinctive feature of a violin body is its hourglass-like shape (narrowed at the middle) and the arching of its top and back. The hourglass shape comprises two upper bouts, two lower bouts, and two concave C-bouts at the waist, providing clearance for the bow. The violin's body is a sound box made from wood, the thinner the better: spruce for the front, maple for the back, the neck and the ribs[2]. While most violins have two-piece backs that are joined together, one-piece backs are preferred due to their increased resonance.

Lesson Six
Stringed Instruments

F Hole: After the vibration from the strings reverberates within the body of the violin, the sound waves are directed out of the body through the F holes.

Bridge: The bridge of the violin is the precise-cut maple piece, forming the lower anchor point of the vibrating length of the strings. It also transmits the vibration of the strings to the body of the instrument when bowing the strings. Its top curve holds the strings at the proper height from the fingerboard in an arc, allowing each to be sounded separately by the bow.

front and back views of violin bridge

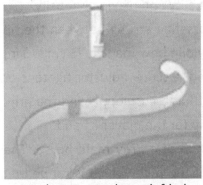
sound post seen through f-hole

Sound post: The sound post, or soul post, fits precisely inside the instrument between the back and top, at a carefully chosen spot near the treble foot of the bridge, which it helps support. It also influences the modes of vibration of the top and the back of the instrument.

Fine tuner: (also called fine adjusters) Fine tuners can be found either on all four strings, or just the E string. Most fine tuners consist of a metal screw that moves a lever attached to the string end. They permit very small pitch adjustments much more easily than the pegs. By turning one clockwise, the pitch becomes sharper (as the string is under more tension), and turning one counterclockwise, the pitch becomes flatter (as the string is under less tension).

Tailpiece / Endpin: The tailpiece is what the strings are attached to at the bottom of the instrument, closest to the players chin. The tailpiece anchors the strings to the lower bout of the violin by means of the tailgut, which loops around an ebony button called the tailpin (sometimes confusingly called the endpin), which fits into a tapered hole in the bottom block.

closeup of a violin tailpiece, with a fleur-de-lis

Chin rest: The chin rest is an additional invention that supports the players chin when they are playing the violin.

String: The strings on the violin are tuned G, D, A, E from low to high. The quality of the string makes a considerable difference to the tonal quality produced by the instrument. Traditionally the strings are made from sheep gut (catgut), modern strings use steel, or various synthetic materials, wound with metals, sometimes plated with silver or gold, except for the E strings, which are normally unwound. However, the sheep gut is also favored for specific requirements. Strings usually have a colored silk wrapping at both ends, for identification of the string (e.g., G string, D string, A string or E string) and to provide friction against the pegs. The four strings are tuned with the help of the pegs at the top and sometimes with the fine tuners on the tailpiece. String longevity depends on string quality and playing intensity.

Bow: The bow is a wooden stick with a ribbon of horsehair strung between the tip and frog (or nut, or heel) at opposite ends. A typical violin bow may be 75 cm (30 in) overall, and weigh about 60 g (2.1 oz). At the frog end, a screw adjuster tightens or loosens the hair. Just forward of the frog, a leather thumb cushion, called the grip, and winding protect the stick and provide a strong grip for the player's hand.

The hair of the bow traditionally comes from the tail of a grey male horse. The natural texture of the horsehair and the stickiness of the rosin which is rubbed on the horsehair help the bow to "grip" the string, and thus when the bow is drawn over the string, the bow causes the string to sound a pitch. The bow sticks are made of costly timbers like snakewood or brazilwood, or inexpensive synthetic materials, as fiberglass.

II. Development

There aren't complete records left today on the origin of the instrument. The direct ancestor of all European bowed instruments is the Arabic rebab[3], which developed into the Byzantine lyra[4] by the 9th century and later the European rebec[5]. The first makers of violins probably borrowed from various developments of the Byzantine lyra. These included the vielle[6] and the lira da braccio[7]. The violin in its present form emerged in early 16th-century northern Italy. The earliest pictures of violins, in spite of three strings, are seen in northern Italy around 1530. One of the earliest explicit descriptions of the violin, including its tuning, was in the *Epitome musical*[8] by Jambe de Fer, published in Lyon[9] in 1556. By this time, the violin had already begun to spread throughout Europe.

The baroque period witnessed the final transition of the instrument. Between the

16th century and the 18th century, the violin became more and more popular in Europe due to its small size and melodic tone, and the violin making entered its golden age. The most famous violin makers (luthiers) between the 16th century and the 18th century are mainly from these schools: the school of Brescia, the school of Cremona, and the school of Venice. The "Charles IX[10]" made by Andrea Amati[11], is the oldest surviving violin, while the "Messiah[12]" or "Le Messie" (also known as the "Salabue") made by Antonio Stradivari[13] in 1716, is the most famous one. At the end of the baroque period, the violin was established as a solo instrument and many famous musicians composed for it since then, such as J.S. Bach[14] and the virtuoso Niccolo Paganini[15].

During the late 18th to the early 20th Century, the violin had undergone some significant technical changes, which transformed the violin from the baroque style to modern style. The neck was lengthened and tapered and tilted, and the fingerboard was lengthened to offer the instrument an even higher pitch range. The angle between the fingerboard and the body was adjusted to result in slightly higher tension, and the bass bar was strengthened and made thicker to support the added tension and increase the resonance of the instrument. The metal strings gradually replaced the sheepgut (catgut) over the first 20 years of the 20th century. Original shapes of bow was convex, while the present is concave which is easier to handle and capable of more delicate changes in playing. All these changes of the violin have resulted in an increased range and a more penetrating sound.

III. Performing Techniques

Correct posture for playing the violin

The violin is usually held under the chin, and supported by the shoulder in either a standing or sitting position. The standard way of holding the violin is with the left side of the jaw resting on the chinrest of the violin, and supported by the left shoulder, often assisted by a shoulder rest. The jaw and the shoulder must hold the violin firmly enough to allow it to remain stable when the left hand goes from a high position (a high pitched note far up on the fingerboard) to a low one (nearer to the pegbox). Good posture is very important to enhance the quality of the sound and minimize possible muscle and back strain to the player.

Violin finger names

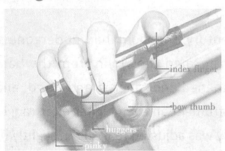
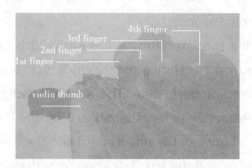

The right hand is known as the bow hand (as this hand controls the bow), and the left is the violin hand, which holds the violin. The fingers on both hands also have different names to differentiate them from one another.

Left hand and pitch production

The left hand determines the sounding length of the string, and thus the pitch of the string, by "stopping" it (pressing it) against the fingerboard with the fingertips, producing different pitches. The player must know exactly where to place the fingers on the strings to play with good intonation (tuning). The main techniques used in the left hand include:

Positions

The placement of the left hand on the fingerboard is called "position". Technically there are 15 positions, in which the first 7 ones are commonly used. When coming to some notes-changing, a change of position is needed, and it is called a "shift of the position[16]".

Open strings

If a string is bowed or plucked without any finger stopping it, it is an open string. This gives a different sound from a stopped string, since the string vibrates more freely at the nut than under a finger. Further, it is impossible to use vibrato fully on an open string.

Double stops, triple stops and chords

Double stopping[17] is when two separate strings are stopped by the fingers and bowed simultaneously, producing two continuous tones (typical intervals include 3rds, 4ths, 5ths, 6ths, and octaves). Double-stops can be indicated in any position, though the widest interval that can be double-stopped naturally in one position is an octave (with the index finger on the lower string and the pinky finger on the higher string). The term "double stop" is often used to encompass sounding an open string alongside a fingered note as well, even though only one finger stops the string.

Because the bow will not naturally strike three strings at once, Where three or four simultaneous notes are indicated, the violinist will typically "split" the chord, choosing the lower one or two notes to play first before immediately continuing onto the upper one or two notes, with the natural resonance of the instrument producing an effect similar to if all four notes had been voiced simultaneously.

Vibrato

Vibrato is a technique of the left hand and arm in which the pitch of a note varies subtly in a pulsating rhythm. While various parts of the hand or arm may be involved in the motion, the end result is a movement of the fingertip bringing about a slight change in vibrating string length, which causes an undulation in pitch.

Vibrato can be produced by a proper combination of finger, wrist and arm motions. One method, called hand vibrato (or wrist vibrato), involves rocking the hand back at the wrist to achieve oscillation. In contrast, another method, arm vibrato, modulates the pitch by movement at the elbow. A combination of these techniques allows a player to produce a large variety of tonal effects. Different types of vibrato will bring different moods to the piece.

Harmonics (also called overtones or partials)

Lightly touching the string with a fingertip, but without fully pressing the string, and then plucking or bowing the string, creates harmonics.

There are two types of harmonics: natural and artificial harmonics (also known as false harmonics). *Natural harmonics* are played on an open string. *Artificial harmonics* are more difficult to produce than natural harmonics, as they involve both stopping the string and playing a harmonic on the stopped note.

Right hand and tone color

The strings may be sounded by drawing the hair of the bow held by the right hand across them (arco) or by plucking them (pizzicato) most often with the right hand.

The right hand, arm, bow and the bow speed are responsible for tone quality, rhythm, dynamics, articulation, and most changes in timbre. The main techniques used in the right hand include:

Bowing techniques

The bow grip is the most essential part of bowing technique. It is usually with the thumb bent in the small area between the frog and the winding of the bow. The other fingers are spread somewhat evenly across the top part of the bow. The pinky finger is curled with the tip of the finger placed on the wood next to the screw. The violin produces louder notes with greater bow speed or more weight on the string. The two methods (greater bow speed or more weigh) are not equivalent, because they produce different timbres; pressing down on the string tends to produce a harsher, more intense sound. One can also achieve a louder sound by placing the bow closer to the bridge.

The sounding point where the bow intersects the string also influences timbre (or "tone colour"). Playing close to the bridge (*sul ponticello*[18]) gives a more intense sound than usual, emphasizing the higher harmonics; and playing with the bow over the end of the fingerboard (*sul tasto*[19]) makes for a delicate, ethereal sound, emphasizing the fundamental frequency.

Various methods of attack with the bow produce different articulations. There are many bowing techniques that allow for every range of playing style. The other main bowing techniques include: legato-style bowing (a smooth, connected, sustained sound suitable for melodies), collé, and a variety of bowings which produce shorter notes, including ricochet, sautillé, martelé, spiccato, and staccato.

Pizzicato (*both left and right*): Pizzicato means pluck or pinch in Italian, it can be delicate, dramatic or percussion. Violinists usually pluck the string with a finger of the *right hand* rather than by bowing. The index finger is most commonly used. Some plucking passages can be no longer than one or two notes, so it is important to practise a fast transition between plucking and bowing. *Left hand* pizzicato is usually done on open strings.

Col legno (Italian for "with the wood"): striking the string(s) with the stick of the bow, rather than by drawing the hair of the bow across the strings. This bowing technique is somewhat rarely used, and results in a muted percussive sound.

Tremolo: It is the very rapid repetition (typically of a single note, but occasionally of multiple notes), usually played at the tip of the bow.

Mute or sordino: Placing a small metal, rubber, leather, or wooden device called a mute, or sordino, to the bridge of the violin gives a softer, more mellow tone, with fewer audible overtones. The mute changes both the loudness and the timbre ("tone colour") of a violin.

IV. Musical Classics

The Four Seasons   Antonio Vivaldi　　　　　（《四季》安东尼奥·维瓦尔第）

Violin Concerto in E Major   Johann Sebastian Bach

　　　　　　　　　　　　　　　（《E 大调小提琴协奏曲》约翰·塞巴斯蒂安·巴赫）

Chaconne   Johann Sebastian Bach

　　　　　　　　　　　　　　　　　　（《恰空舞曲》约翰·塞巴斯蒂安·巴赫）

Eine Kleine Nachtmusik   Wolfgang Amadeus Mozart

　　　　　　　　　　　　　　　　　（《小夜曲》沃尔夫冈·阿玛多伊斯·莫扎特）

Ode to Joy   Ludwig Van Beethoven　　（《欢乐颂》路德维希·凡·贝多芬）

Adagio for Strings   Samuel Barber　　　　（《弦乐慢板》塞缪尔·巴伯）

Meditation   Jules Massenet　　　　　　　　（《沉思》儒勒·马斯奈）

Salut d'Amour   Edward Elgar　　　　　（《爱的礼赞》爱德华·艾尔加）

The Swan   Charles Camille Saint-Saëns　（《天鹅》夏尔·卡米尔·圣-桑）

Skylark   Ralph Vaughan Williams　　　（《云雀》拉尔夫·沃恩·威廉斯）

The Blue Danube   Johann Strauss II　（《蓝色多瑙河》约翰·施特劳斯二世）

Canon in D Major   John Pachelber　　（《D 大调卡农》约翰·帕赫贝尔）

Concerto in E Minor   Jakob Ludwig Felix Mendelssohn Bartholdy

　　　　　　　　（《E 小调协奏曲》雅克布·路德维希·菲利克斯·门德尔松）

Wanderer's Song/Gypsy's Song   Pablo de Sarasate

　　　　　　　　　　　　　（《流浪者之歌》或《吉卜赛之歌》帕布罗·德·萨拉萨蒂）

Swan Lake   Pyotr Ilyich Tchaikovsky　（《天鹅湖》彼得·伊里奇·柴科夫斯基）

The Butterfly Lovers   He Zhanhao&Chen Gang　　（《梁祝》何占豪，陈钢）

Notes on the text

1. **fiddle** 提琴类乐器；小提琴＜非正式＞

2. The violin's body is a sound box made from wood, the thinner the better: spruce for the front, maple for the back, the neck and the ribs. 小提琴的主体是由木头制成的音箱，木头越薄越好：云杉木用来做面板，枫木用来做背板、琴颈和侧板。

3. rebab 雷贝琴（一种起源于阿拉伯的弓弦弹拨乐器，尤用于北非、中东和印度次大陆。）
4. *lyre* （古希腊的）里尔琴 / 七弦竖琴（古代 U 形拨弦乐器）。
5. *rebec* 雷贝克琴（大约在 10 世纪时从阿拉伯传入欧洲的弦乐器。长颈，圆形或梨形琴身，通常有三根弦。）
6. *vielle* 维埃尔琴（6~7 根弦），是阿拉伯人的雷贝琴的一种变种。
7. *lira da braccio* 高音里拉琴，出现在大约 15 世纪。
8. the Epitome musical 音乐摘要
9. Lyon 里昂，法国东南部城市。
10. Charles IX: 查理九世，现存最早的小提琴，由安德里亚·阿玛蒂在 1564 年制作于意大利北部城市克雷莫纳（Cremona）。
11. Andrea Amati 安德里亚·阿玛蒂（1505—1577），意大利北部克里莫纳的一位著名制琴师，被认为是小提琴的发明者。
12. Messiah 弥赛亚（Le Messie），也作 Salabue，最有名的小提琴，是安东尼奥·斯特拉迪瓦里（Antonio Stradivari）1716 年制作，这把小提琴现藏于英国牛津的 Ashmolean 博物馆。
13. Antonio Stradivari 安东尼奥·斯特拉迪瓦里（1644—1737），意大利的弦乐器制作师，这个职业中最伟大的一位成员。
14. J.S. Bach 约翰·塞巴斯蒂安·巴赫（1685—1750），出生于德国图林根州的埃森纳赫，巴洛克时期德国作曲家、键盘演奏家，被称为"西方音乐之父"。
15. Niccolo Paganini 尼科罗·帕格尼尼（1782—1840），意大利小提琴 / 吉他演奏家、作曲家，是历史上最著名的小提琴演奏大师之一。
16. shift of the position 换把
17. double stopping 双音演奏
18. sul ponticello 靠近琴马演奏
19. sul tasto 靠近指板演奏

Terms

luthier	/ˈlutiə/	琴师
bowed stringed instrument	/boʊd ˈstrɪŋd ˈɪnstrəmənt/	弓弦乐器
scroll	/skrəʊl/	琴头
tuning peg /pegbox	/ˈtuːnɪŋ peg/, /peg bɑːks/	弦轴 / 弦轴箱
fingerboard	/ˈfɪŋɡərbɔːrd/	指板
tailpiece / endpin	/ˈteɪlpiːs/, /endpɪn/	尾柱

Lesson Six
Stringed Instruments

screw	/skruː/	琴钉
bridge	/brɪdʒ/	琴马
sound post	/saʊnd pəʊst/	音柱
bass bar	/beɪs bɑːr/	音梁；低音梁
chin rest	/tʃɪn rest/	腮托
timbre	/ˈtæmbər/	音色
pizzicato/ pluck	/ˌpɪtsɪˈkɑːtəʊ/, /plʌk/	拨弦
vibrato	/vɪˈbrɑːtəʊ/	揉弦
position	/pəˈzɪʃn/	把位
harmonics	/hɑːrˈmɑːnɪks/	泛音
perfect fifth	/ˈpɜːrfɪkt fɪfθ/	纯五度
arpeggio	/ɑːrˈpedʒioʊ/	琶音
mute (or sordino)	/mjuːt/	弱音器
tremolo	/ˈtreməloʊ/	震音；颤音
bowing	/ˈboʊɪŋ/	（法）运弓
detached	/dɪˈtætʃt/	分弓
spiccato	/spɪˈkɑːtoʊ/	跳弓
legato	/lɪˈgɑːtəʊ/	圆滑连弓
staccato	/stəˈkɑːtəʊ/	顿弓
bow grip	/baʊ grɪp/	握弓
col legno		敲弓

Exercises

I. Comprehension questions

1. What is the status of the violin in the musical instruments?
2. Which members are there in the violin family?
3. Which period witnessed the final transition of the violin in Western musical history?
4. What is the function of the sound post?

5. What is the main difference of the left and the right hand in the performing technique when playing the violin?

II. Translating useful expressions

1. "小提琴"一词是"弦乐"的意思，小提琴属弓弦乐器。
2. 小提琴一般有四根弦，从低音到高音分别为 G 弦、D 弦、A 弦和 E 弦。
3. 左手常用的演奏技巧有把位、拨弦、揉弦和泛音。
4. The distinctive features of body of the violin are the hour-glass shape, hollow, and with two f-shaped soundholes.
5. The right hand, arm, bow and the bow speed are responsible for tone quality, rhythm, dynamics, articulation, and most changes in timbre.
6. The range of the violin is four octaves, and it is often possible to play higher, depending on the length of the fingerboard and the skill of the violinist.

III. Brainstorm

The first violin, as the leader of a symphony orchestra, occupies an important position in the symphony orchestra, but only a virtuoso chief violin doesn't make a good symphony orchestra. *One strand of silk doesn't make a thread; one tree doesn't make a forest.* (单丝不成线，独木不成林。) Give a specific example to brainstorm the importance of collaboration.

Viola

The viola is a string instrument that is bowed, plucked or played with varying techniques. The name originates from the Italian term *"viola da braccio"*, but in French the instrument is called *alto*, and in Germany *Bratsche*. The viola is slightly larger than a violin with a lower and deeper sound. It often plays the "inner voices" in string quartets and symphonic writing, and occasionally plays a major, soloistic role in orchestral music. Unlike most other instruments, the music for the viola typically uses the alto clef, with occasionally treble clef when needed. The person who plays the viola is called a violist or a viola player.

I. Construction

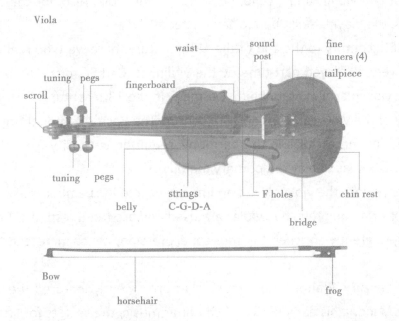

The viola is similar in material and construction to the violin. It does not have a standard full size, the average length is 41 cm (16 in). Small violas made for children are equivalent to half-size violins.

Recent innovations make the violas shorter and lighter, associating the ergonomic problems with keeping the traditional sound.

Compared with violins, the viola has a heavier bow with a wider band of horsehair, and the outside corner of the bow frog (or heel in the UK) is more rounded.

One of the most notable makers of violas in the 20th century was English-born

Australian A. E. Smith, whose violas are sought after and highly valued.

II. Development

As for what the viola was originated from and how it was invented, two instruments are usually mentioned. The 1st one is the Italian *viola da braccio* (of the arm), in distinction to the viola da gamba (of the leg) and the viol family, which at first seemed to encompass the whole violin family, eventually came to refer to the alto member specifically. The 2nd one is the viola d' amore, an alto instrument played like a violin on the shoulder, which was the direct predecessor of the viola in the orchestra. Bach used the instrument in several of his important mass pieces.

Prior to the 18th century, the viola was occasionally used, mainly to strengthen the bass line or fill in the harmony. At that time the most popular chamber string trio consisted of two violins and a cello, no score for the viola. Also, for the sake of the varied sizes, the playing skills were often unsteady.

From the 16th century to the end of the 19th century there were no real violists, for the violas were all played part-time by the violinists. Carl Stamitz[1] (1745-1801) was the earliest viola player in Mannheim, Germany in the 18th century

The heyday of the viola came in the 18th century.The Bach's Brandenburg Concertos No.3 scored for three violas in the harmony, requiring virtuosity from the violists; No.6 used two violas to play the primary melody.

In the same period, the viola played an important role in chamber music as well. In Mozart's six string quintets, two violas are used and are freed (esp. the 1st viola) for solo passages, greatly enriched the musical expression, hence increased the variety of writing.

In the 19th century, Brahms wrote music that prominently featured the viola in his early works. Among his early pieces of chamber music, the sextets for strings Op.18 and Op. 36 contain what amounts to solo parts for both violas.

In the 20th century, encouraged by the emergence of specialized soloist, such as Lionel Tertis[2] and William Primrose[3], more composers began to write for the viola, including a substantial amount of chamber and concert works.

III. Tuning

The viola is the middle or alto voice of the violin family, its tuning is a perfect fifth below the violin, and an octave above the cello. The four strings are tuned in fifths, from low to high: C-G-D-A. The C string is tuned an octave below middle C.

Besides C-G-D-A, other tunings (*scordatura*[4]) are occasionally employed both in classical music and some folk styles. In Mozart's *Sinfonia Concertante for Violin, Viola and Orchestra in Eb*, the viola part was in D major, and was specified to raise the strings in pitch by a semitone, giving the viola a brighter tone so the rest of the ensemble wouldn't overpower it.

IV. Playing Techniques

Compared with the violin, the playing technique of viola bears certain differences, for the sake of: larger size, less responsive strings, heavier bow and the notes spreading out farther along the fingerboard.

Due to its larger size, a violist must bring the left elbow farther forward or around, so as to reach the lowest string, which allows the fingers to press firmly and thus creates a clearer tone. Also the right arm must hold the bow farther away from the player's body.

A violist uses wider-spaced fingerings. They use a wider and more intense vibrato in the left hand, using the fleshier pad of the finger rather than the tip. Different positions are often used, including half position.

The viola is strung with thicker strings, thus it responds to changes in bowing more slowly, to which the solution is either beginning to move the bow sooner, or bow harder to make the strings vibrate.

V. Musical Classics

Brandenburg Concertos J. S. Bach
(《布兰登堡协奏曲》约翰·塞巴斯蒂安·巴赫)

String Quintets Wolfgang Amadeus Mozart
(《弦乐五重奏》沃夫冈·阿玛多伊斯·莫扎特)

Sinfonia Concertante Wolfgang Amadeus Mozart
(《协奏曲序曲》沃夫冈·阿玛多伊斯·莫扎特)

Viola Sonata in C Minor Felix Mendelssohn
(《C小调中提琴奏鸣曲》菲力克斯·门德尔松)

Fairy Tales Storybook Robert Schumann   (《童话绘本》罗伯特·舒伯特)

Sextets for Strings Op.18&36 Johannes Brahms
（《弦乐六重奏 第 18、36 号》约翰内斯·勃拉姆斯）
Two Songs for Alto with Viola and Piano Johannes Brahms
（《两首中音作品：中提琴与钢琴》约翰内斯·勃拉姆斯）
Concerto for Viola and Orchestra Bela Bartok
（《中提琴协奏曲》贝拉·巴托克）

Notes on the text

1. Carl Stamitz 卡尔·斯塔米茨，德国作曲家、小提琴家。
2. Lionel Tertis 昂内尔·特蒂斯，英国中提琴家，被称为现代中提琴之父。
3. William Primrose 威廉·普利姆罗斯，苏格兰中提琴家，被认为是中提琴演奏家中最杰出代表。
4. Scordatura 特殊调弦法

Terms

viola	/viˈəʊlə/	中提琴
braccio	/ˈbrɑːtʃoʊ/	（意）手臂
heyday	/ˈheɪdeɪ/	全盛时期
alto clef	/ˈæltəʊ klef/	中音谱号
treble clef	/ˈtrebl klef/	高音谱号
bass clef	/beɪs klef/	低音谱号
quartet	/kwɔːrˈtet/	四重奏
quintet	/kwɪnˈtet/	五重奏
sextet	/seksˈtet/	六重奏
substantial	/səbˈstænʃl/	大量的
form	/fɔːrm/	形制
violist	/viˈəʊlɪst/	中提琴手
fingering	/ˈfɪŋgərɪŋ/	指法
bowing	/ˈboʊɪŋ/	弓法
position	/pəˈzɪʃn/	把位
tuning	/ˈtuːnɪŋ/	调弦
scordatura	/ˌskɔːdɑːˈtuːrɑː/	（意）特殊调弦
concerto	/kənˈtʃɜːrtoʊ/	协奏曲
concerto grosso	/kənˈtʃɜːrtoʊ ˈgroʊsoʊ/	大协奏曲
chamber music	/ˈtʃeɪmbər ˈmjuːzɪk/	室内乐

Lesson Six
Stringed Instruments

Exercises

I. Comprehension questions

1. How to call the person who plays the viola?
2. Which instrument is the direct forerunner of the viola in the orchestra?
3. How is the viola different from the violin?
4. In which register does the viola play in the violin family?
5. How to hold a viola when performing?

II. Translating useful expressions

1. 中提琴是弓弦乐器，是小提琴家族中的中音乐器。
2. 中提琴整体调律比小提琴低五度，比大提琴高八度。
3. 中提琴演奏的指法和弓法和小提琴相似。
4. The four strings are tuned in fifths, from low to high: C-G-D-A, the C string is tuned an octave below middle C.
5. In Mozart's *Sinfonia Concertante for Violin, Viola and Orchestra in Eb*, the viola part was in D major, and was specified to raise the strings in pitch by a semitone, giving the viola a brighter tone so the rest of the ensemble wouldn't overpower it.
6. Different positions are often used, including half position.

III. Brainstorm

The viola used not to be taken seriously, yet it has made its way in the orchestra and been loved by many people. How would that inspire you in the way of your career making or life building?

Cello

The cello (or *violoncello*) is a bowed string instrument of the violin family, which is known for its rich and warm tone color. The cello is good at playing lyrical melodies and expressing deep and complex feelings, and is acclaimed as "the lady of music". The four strings of the cello are usually tuned in perfect fifths and its music is written in the bass clef. Known for its warm and rich timbre, many famous musicians write works for the cello, and it is an indispensable bass or tenor string instrument in an orchestra. The earliest surviving cellos are made by Andrea Amati[1], the first known member of the celebrated Amati family of luthiers. One of the most famous cellos is the Davidov[2] Stradivarius, which is currently possessed by Yo-Yo Ma, a well-known cellist.

The cello plays an important role in the standard symphony orchestra, which usually includes eight to twelve cellists. Much of the time, cellos provide part of the low-register harmony for the orchestra. The cello section, in a standard symphony orchestra, is located on the stage left in the front, opposite the first violin section. The principal cellist who always sits closer to the audience, is the section leader, determining bowing for the section in combination with other string principals, playing solos, and leading entrances(when the section begins to play its part). Nowadays, cellos can not only be used in symphony orchestra, but also in solo, ensembles, popular music, jazz, world music and even in modern Chinese orchestra.

I. Construction

The construction of the cello is closely identical to the other members of the violin family, except for the endpins.

The tailpiece and endpin are found in the lower part of the cello. The tailpiece is the part of the cello to which the "ball ends" of the strings are attached by passing them through holes. The endpin or spike is made of wood, metal, or rigid carbon fiber and supports the cello in playing position. Modern endpins are retractable and adjustable.

Lesson Six
Stringed Instruments

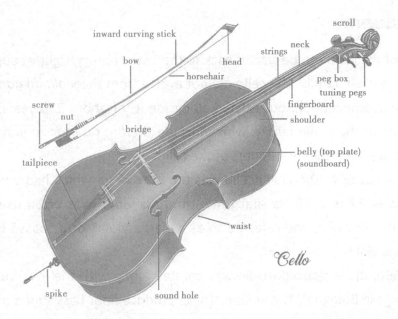

Cello

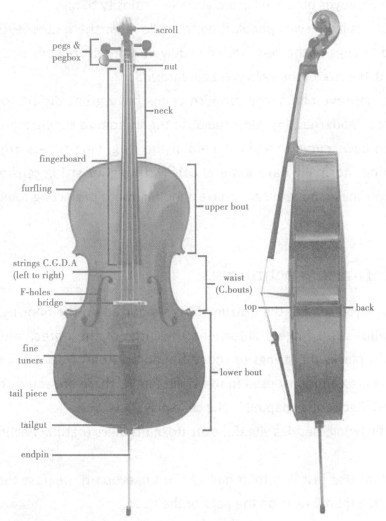

II. Development

The origin of the cello can be traced back to the 16th century, tightly coupled to the history of the violin family. The cello did not evolve from the *viola da gamba*, which is a general misconception, but existed alongside it for about 250 years. The likely predecessors of the violin family, which include the *lire da braccio* and the *rebec*, could also have been the forerunners of the cello.

The direct ancestor to the violoncello was the *bass violin* which had possibly been invented as early as 1538. At that time, the bass violin was often used to be in consort with the violin, and referred to as a "large viola" or a "basso de *viola da braccio*"(bass viola).

 Around 1660, the violoncello/cello was created along with the invention of wire-wound strings in Bologna[3]. The wound strings produce finer bass sound on a smaller body, which improves both the timbre and the virtuosity to play.

Around 1700, Italian players popularized the cello in northern Europe, but the sizes, names, and tunings of the cello varied widely in different regions and times. Until around 1750, the size of the cello was standardized.

After many improvements, the modern cellos have great differences from the baroque ones. Modern cellos add endpins at the bottom to support the instrument, and modern bows curve in and are held at the frog. Fine-tuners are also added and the strings normally have a metal core with higher string tension. All these improvements make modern cellos have a louder, more projecting tone, with fewer overtones.

III. Performing Techniques

The cello is played seated, the instrument is supported on the floor by the endpin. Still, the cellist's left fingers determine the pitch of the notes, while the right hand bows or plucks the strings to sound the notes. Apart from the fingerings and bowings that are commonly used in the violin family, there are some specific terms that are more frequently adapted in the cello playing.

In terms of fingering, besides vibrato, pizzcato, harmonics (natural / artifitial)... there are:

Neck position: Use just less than half of the fingerboard, nearest the top of the instrument; the thumb rests on the back of the neck.

Thumb position: A general name for notes on the remainder of the fingerboard; the side of the thumb is used to play notes.

Glissando/sliding: An effect played by sliding the finger up and or down the fingerboard without releasing the string; causes the pitch to rise and fall smoothly without separate, discernible steps.

In terms of bowing which determines the tone production and volume of sound, there are 4 most important factors that influence the quality of the tone. They are: the *weight* applied to the string, the *angle* of the bow on the string, bow *speed* and the *point of contact* of the bow hair with the string (abbr. WASP).

Other bowing techniques include: staccato, legato, col legno, spiccato.

IV. Musical Classics

Six Unaccompanied Cello Suite   Johann Sebastian Bach
　　　　　　　　（《六首无伴奏大提琴组曲》约翰·塞巴斯蒂安·巴赫）

Cello Concerto in D Major No. 2   Joseph Haydn
　　　　　　　　（《D大调第二号大提琴协奏曲》约瑟夫·海顿）

Cello Sonatas No.3 and No.5   Ludwig van Beethoven
　　　　　　　　（《第三、第五号大提琴奏鸣曲》路德维希·凡·贝多芬）

Cello Concerto in A Minor   Robert Schumann
　　　　　　　　（《a小调大提琴协奏曲》罗伯特·舒曼）

Cello Concerto No.1 and No.2   Johannes Brahms
　　　　　　　　（《第一、第二号大提琴奏鸣曲》约翰内斯·勃拉姆斯）

Cello Concerto in E Minor   Edward Elgar
　　　　　　　　（《E小调大提琴协奏曲》爱德华·埃尔加）

Notes on the text

1. Andrea Amati 安德雷亚·阿马蒂，意大利16世纪著名制琴师。

2. Davidov Stradivarius 大卫朵夫·斯特拉迪瓦里，著名大提琴，被柴可夫斯基称为"大提琴界的沙皇"的俄国大提琴家大卫朵夫曾经拥有，由意大利著名制琴师斯特拉迪瓦里制作，现为华人提琴家马友友所收藏。

3. Bologna 博洛尼亚，意大利城市。

Terms

scroll	/skrəʊl/	琴头
tuning peg	/ˈtuːnɪŋ peg/	弦轴
neck	/nek/	琴颈
nut	/nʌt/	琴枕；弦枕
string	/strɪŋ/	琴弦
fingerboard	/ˈfɪŋgərbɔːrd/	指板
body	/ˈbɑːdi/	琴身
F Hole	/ef həʊl/	孔；音孔
bridge	/brɪdʒ/	琴马
tailpiece	/ˈteɪlpiːs/	系弦板
endpin	/end pɪn/	琴脚
pizzicato	/ˌpɪtsɪˈkɑːtəʊ/	拨弦
vibrato	/vɪˈbrɑːtəʊ/	颤音
harmonics	/harˈmɑnɪks/	泛音
glissando/ sliding	/glɪˈsændoʊ/, /ˈslaɪdɪŋ/	滑音
staccato	/stəˈkɑːtoʊ/	顿弓
legato	/lɪˈgɑːtəʊ/	圆滑连弓
col legno		敲弓
spiccato	/spɪˈkɑːtoʊ/	跳弓

Exercises

I. Comprehension questions

1. What kind of melodies and feelings is the cello good at expressing?
2. How are the four strings tuned?
3. Where is the cello section located?
4. Did the cello evolve from the *viola da gamba*?
5. What is the direct ancestor to the violoncello?

II. Translating useful expressions

1. 大提琴是弓弦乐器，是提琴家族的低音乐器。
2. 大提琴五度调弦，通常使用低音谱号记谱。
3. 乐团大提琴首席是大提琴声部的领导者。
4. The cello plays an important role in the standard symphony orchestra, which usually includes eight to twelve cellists. Much of the time, cellos provide part of the low-register harmony for the orchestra.
5. The wound strings produce finer bass sound on a smaller body, which improves both the timbre and the virtuosity to play.
6. In terms of bowing which determines the tone production and volume of sound, there are 4 most important factors that influence the quality of the tone. They are: the *weight* applied to the string, the *angle* of the bow on the string, bow *speed* and the *point of contact* of the bow hair with the string (abbr. WASP).

III. Brainstorm

Nowadays the cello has been a regular member of many standard Chinese traditional orchestras. Please explain the reasons from the professional perspectives, and then brainstorm your understanding of the tolerance of Chinese culture.

Double Bass

The double bass is also known as the bass. It's the largest and the lowest-pitched bowed or plucked string instrument in the modern symphony orchestra. The person who plays the double bass is called the "bassist", "double bassist", "double bass player", "contrabassist", or "contrabass player". The double bass is a transposing instrument whose music is notated one octave higher than tuned to avoid excessive ledger lines below the staff.

A standard orchestra would have the double bass as one member of its string section, typically with 8 basses performed in unison, while in smaller orchestras may only have 4 basses and exceptional cases 10. The bass is widely featured in western classical genres, eg. concerto, solo and chamber music. It is also used in modern varieties, such as jazz, blues, rock and roll, bluegrass[1], rockabilly[2], psychobilly[3], country music and folk music.

I. Construction

The double bass stands around 180cm, other sizes such as a 1/2 or 3/4, accommodating a player's height and hand size. The instrument is typically made from several types of wood: basically maple for the back, spruce for the top, and ebony for the fingerboard.

The double bass is closest in construction to violins, but also has some notable similarities to the violone. The major two approaches used in the outline design are: the violin form and the viola da gamba form. The former has two pairs of violin corners and a round, carved back, while the latter with no violin corners and having a flat and angled back similar to the viol family. A third approach is called the busetto[4] shape, which is less-commonly employed. Unlike the bulging shoulders that the violin family has, the double bass features the sloped shoulders for the convenience of playing the upper range.

The double bass bow comes in two distinct forms. The French, or "overhand" bow is similar in shape and implementation to the bow of the violin family. The German, or "Butler" bow, typically broader and shorter, the design and the holding manner descending from the older viol instrument family. The bow hair chooses either white or black horsehair or a combination of the two (known as "salt and pepper").

the violin form   *the viola da gamba form*   *the busetto form*

II. Development

The double bass is generally regarded as a modern descendant of the string family of instruments that originated in Europe in the 15th century. Its exact lineage is still in some debate. The experts' controversy is mainly between the violin family and the viol family, for the obvious features in existing instruments inherited from both sides.

During the 18th century, the double bass enjoyed a period of popularity, many of the most well-known composers of that era wrote pieces for it. For instances: Haydn composed solo passages in some of his symphonies, but grouping bass and cello parts together; Beethoven paved the way for separate double bass parts which became more common in the romantic era; Brahms, whose father was a double bass player, wrote many difficult and prominent parts in his symphonies, which are often chose to be the pieces for orchestra auditions today.

The number of the strings of the bass has not always been standard. Before the 20th century, many double basses only had 3 strings. Nowadays we can still find double basses with strings of 3—6.

III. Tuning

For the sufficient holding required, the double bass are fitted with the machine tuners rather than the wooden friction pegs used in the rest of the violin family.

The double bass is the only instrument that is tuned in fourths in the violin family, the others being tuned in fifths. The standard tuning is E-A-D-G from low to high, which is also the tuning in orchestra. However in classical solo playing the bass is usually tuned a whole tone higher F#-B-E-A, which is called "solo tuning".

Throughout classical repertoire, there are notes that fall below the range of a standard bass. To make this happen, there are some methods. Bassists may play the notes below E an octave higher, if sounded awkward, the entire passage may be transposed up an octave. Also, the players could tune the low E string down to the lowest note required in the piece: usually D or C. A third option could be fitting up a "low-C extension"—an extra section of fingerboard, extending the fingerboard and giving additional 4 semitones downwards to low C or even low B. Or the player may employ a 5-string bass, with the additional lower string tuned to C or B (more commonly in modern times). Several major European orchestras use basses with a fifth string.

Beside the standard four-stringed instrument, there are basses with other numbers of strings, together with their common tunings are as following: 3-stringed / A-D-G; 5-stringed / B-E-A-D-G or E-A-D-G-C depending on either additional string is added; 6-stringed/ B-E-A-D-G-C which is ideal for solo and orchestral playing with a more playable range.

IV. Postures

The double basses are played by either standing or sitting. Bassists who stand and bow sometimes set the endpin by aligning the 1st finger in either first or half position with eye level, while players who sit generally use a stool about the height of the player's trousers inseam length. They stand, or sit on a high stool, and lean the instrument against their body, turned slightly inward to put the strings comfortably in reach. The stance is a key reason for the bass's sloped shoulders—the narrower shoulders facilitate playing the strings in their higher registers.

When playing the instrument's upper range, the player shifts the hand from behind the neck, flattens it out, using the side of the thumb to press down the string. The technique is called "thumb position", and it is also used on the cello.

V. Performing Styles

The playing style of the basses is discussed in classical and modern genres.

In classical pieces, the playing is mainly focused on performing with the bow (arco), and producing a good bowed tone. The player can either use a bow traditionally or strike the wood of the bow against the string. The common-used bow articulations include: detache, legato, staccato, sforzato, martele, tremolo, spiccato, sul ponticello... some of these articulations can be combined. In some orchestral repertoires and tango music, both arco and pizzicato are employed. But the pizzicato in these parts generally require simple notes (quarter notes, half notes, whole notes), rather than rapid passages.

In jazz, rockabilly and blues, pizzicato is the norm. The bassists develop virtuoso pizzicato techniques to play rapid solos that incorporate fast-moving triplet and 16th note figures. Besides, "ghost notes" are frequently improvised into bass lines, to add to the rhythmic feel and add fills to a bass line. However, some solos and occasional written parts in modern jazz also call for bowing.

Notes on the text

1. bluegrass 蓝草音乐，乡村音乐的另一个分支。
2. Rockabilly 山区乡村摇滚，摇滚乐最早的形式之一。
3. Psychobilly 精神摇滚
4. Busetto 布塞托，意大利北部小城，是伟大的作曲家威尔第的家乡。

Terms

contrabass	/ˈkɑːntrəbeɪs/	（倍）低音的
bassist	/ˈbeɪsɪst/	低音提琴手
notation	/nəʊˈteɪʃn/	记谱
transposing	/trænˈspəʊzɪŋ/	移 / 转调的
debate	/dɪˈbeɪt/	争论
descendant	/dɪˈsendənt/	后代；派生物
sloped	/sləʊpt/	倾斜的
bulging	/ˈbʌldʒɪŋ/	鼓起的
employ	/ɪmˈplɔɪ/	使用
quarter/ half/ whole note	/ˈkwɔːrtər nəʊt/, /hæf nəʊt/, /həʊl nəʊt/	四分 / 二分 / 全音符
norm	/nɔːrm/	常态
triplet	/ˈtrɪplət/	三连音
recital	/rɪˈsaɪtl/	独奏 / 唱音乐会
repertoire	/ˈrepərtwɑːr/	保留 / 全部曲目
stool	/stuːl/	高脚凳
liberal arts	/ˈlɪbərəl ˈɑːrts/	人文学科
credential	/krəˈdenʃl/	文凭；资格
articulation	/ɑːrˌtɪkjuˈleɪʃn/	发音

Exercises

I. Comprehension questions

1. How many basses are scored in a standard orchestra?
2. Regarding the lineage of the double bass, what's the experts' controversy?
3. What is the common design for the bass outline?
4. Why are some bass bows called "salt and pepper"?
5. How is the bass special in the violin family in terms of tuning?

II. Translating the useful expressions

1. 低音提琴是在现代交响乐团中体积最大、音高最低的弓／拨弦乐器。
2. 低音提琴既可以站着演奏，也可以坐着演奏。
3. 低音提琴的标准定弦从低到高为：E-A-D-G。
4. The double bass is a transposing instrument whose music is notated one octave higher than tuned to avoid excessive ledger lines below the staff.
5. Haydn composed solo passages in some of his symphonies, but grouping bass and cello parts together.
6. Unlike the bulging shoulders that the violin family have, the double bass features the sloped shoulders for the convenience of playing the upper range.

III. Brainstorm

In Camille Saint-Saens' *The Carnival of the Animals*, the double bass is portrayed as "the elephant". Please discuss the resemblance of the two with your classmates.

Classical Guitar

The classical guitar (also known as the nylon-string guitar or Spanish guitar) is a plucked-string instrument, and is a member of the guitar family used in classical music and other styles. The classical guitar, an acoustic wooden string instrument with strings made of nylon or gut, is a precursor of the modern acoustic[1] and electric guitars, both of which use metal strings.

A guitar family mainly includes classical guitar, steel-string acoustic guitar, electric guitar, flamenco guitar[2], and bass guitar, in which the flamenco guitar derives from the modern classical guitar, but has differences in material, construction and sound. Today's modern classical guitar was established by the late designs of the 19th-century Spanish luthier, Antonio Torres Jurado[3].

The classical guitar is distinguished by a number of characteristics:

1. It is an acoustic instrument. The sound of the plucked string is amplified by the soundboard and resonant cavity of the guitar.
2. It has six strings, though some classical guitars have seven or more strings.
3. All six strings are made from nylon, or nylon wrapped with metal, as opposed to the metal strings found on other acoustic guitars. Nylon strings also have a much lower tension than steel strings, as do the predecessors to nylon strings, gut strings (made from ox or sheep gut). The lower three strings ('bass strings') are wound with metal, commonly silver-plated copper.
4. Because of the low string tension: The neck can be made entirely of wood without a steel truss rod, so the interior bracing can be lighter.
5. Typical modern six-string classical guitars are 48–54 mm wide at the nut, compared to around 42 mm for electric guitars.
6. Classical fingerboards are normally flat and without inlaid fret markers, or just have dot inlays on the side of the neck—steel string fingerboards usually have a slight radius and inlays.
7. Classical guitarists use their right hand to pluck the strings. Players shape their fingernails for ideal tone and feel against the strings.
8. Strumming is a less common technique in classical guitar, and is often referred to by the Spanish term "rasgueo", or for strumming patterns "rasgueado[4]", and uses the backs of the fingernails. Rasgueado is integral to Flamenco guitar.
9. Machine heads at the headstock of a classical guitar point backwards—in contrast

to most steel-string guitars, which have machine heads that point outward. The overall design of a Classical Guitar is very similar to the slightly lighter and smaller Flamenco guitar.

I. Construction

The classical guitar is mainly composed of headstock, nut, machine heads (or pegheads, tuning keys, tuning machines, tuners), frets, neck, heel, body, bridge, bottom deck, soundboard, body sides, sound hole with rosette inlay, strings, saddle (bridge nut), and fretboard.

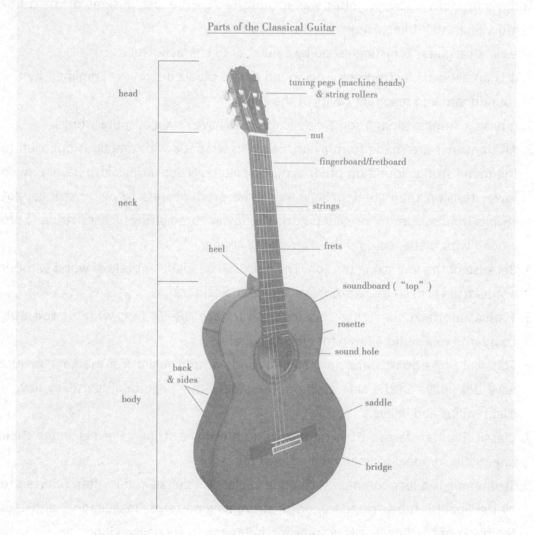
Parts of the Classical Guitar

Lesson Six
Stringed Instruments

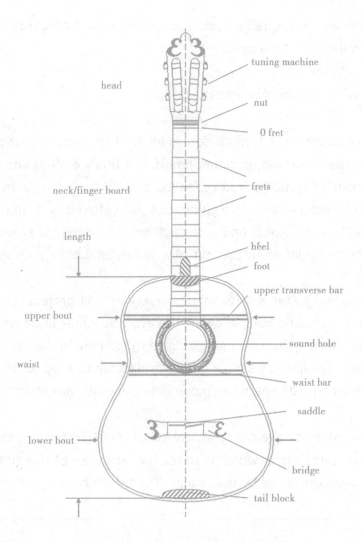

Fretboard

The fretboard (also called the fingerboard) is a piece of wood embedded with metal frets that constitutes the top of the neck. It is flat or slightly curved. Fretboards are most commonly made of ebony, but may also be made of rosewood, some other hardwood, or of phenolic composite.

Fret

Frets are the metal strips (usually nickel alloy or stainless steel) embedded along the fingerboard and placed at points that divide the length of string mathematically. The strings' vibrating length is determined when the strings are pressed down behind the frets. Each fret produces a different pitch and each pitch spaced a half-step apart on the 12 tone scale. Every twelve frets represent one octave. This arrangement of frets results in equal tempered tuning.

Neck

A classical guitar's frets, fretboard, tuners, headstock, all attached to a long wooden extension, collectively constitute its neck. The wood for the fretboard usually differs from the wood in the rest of the neck.

Neck joint or "heel"
This is the point where the neck meets the body.

Body
The body of the instrument is a major determinant of the overall sound variety for acoustic guitars. The guitar *top*, or soundboard, is a finely crafted and engineered element often made of spruce or red cedar. Considered the most prominent factor in determining the sound quality of a guitar, this thin (often 2 or 3 mm thick) piece of wood has a uniform thickness and is strengthened by different types of internal bracing. The *back* and *sides* are made out of a variety of woods such as mahogany and maple etc.

The body of a classical guitar is a resonating chamber that projects the vibrations of the body through a sound hole, allowing the acoustic guitar to be heard without amplification. The sound hole is normally a single round hole in the top of the guitar (under the strings), though some have different placement, shapes, or numbers of holes. How much air an instrument can move determines its maximum volume.

Bridge
The main purpose of the bridge on a classical guitar is to transfer the vibration from the strings to the soundboard, which vibrates the air inside of the guitar, thereby amplifying the sound produced by the strings. The bridge holds the strings in place on the body.

II. Development

The evolution of the classical guitar and its repertoire spans more than four centuries. It has a history that was shaped by contributions from earlier instruments, such as the lute, the vihuela[5], and the baroque guitar.

Origin
The origins of the modern guitar are not certainly known. Guitar-like instruments appear in ancient carvings and statues recovered from Egyptian, Sumerian, and Babylonian civilizations. During the late Middle Ages, gitterns called "guitars" were in use, but their construction and tuning was different from modern guitars. The guitarra latina (拉丁吉他) in Spain had curved sides and a single hole. The guitarra morisca (摩尔吉他), which appears to have had Moorish influences, had an oval soundbox and many sound holes on its soundboard.

Guitarra Latina (left) and Guitarra Morisca (right)

Renaissance guitar

Alonso de Mudarra's[6] book *Tres Libros de Música*, published in Spain in 1546, contains the earliest known written pieces for a four-course guitarra. This four-course "guitar" was popular in France, Spain, and Italy.

Vihuela

By the 15th century, a four-course double-string instrument called the *vihuela de mano*, that had tuning like the later modern guitar except on one string and similar construction, first appeared in Spain and spread to France and Italy. Instead, the lute-like vihuela appeared with two more strings that gave it more range and complexity. In its most developed form, the vihuela was a guitar-like instrument with six double strings made of gut, tuned like a modern classical guitar with the exception of the third string, which was tuned half a step lower. It has a high sound and is rather large to hold. Few have survived and most of what is known today come from diagrams and paintings.

Baroque guitar, "early romantic guitar" or "guitar during the classical music era"

In the middle of the 16th century, influences from the vihuela and the Renaissance guitar were combined and the baroque five-string guitar appeared in Spain. The baroque guitar quickly superseded the vihuela in popularity in Spain, France and Italy. In the late 18th century the six-string guitar quickly became popular at the expense of the five-string guitars.

Modern classical guitar

The modern classical guitar (also known as the "Spanish guitar"), the immediate forerunner of today's guitars, was developed in the 19th century by Antonio de Torres Jurado, Ignacio Fleta, Hermann Hauser Sr., and Robert Bouchet. During the 19th century the Spanish luthier and player Antonio de Torres Jurado[8] gave the modern classical guitar its definitive form, with a broadened body, increased waist curve, thinned belly, improved internal bracing. The modern classical guitar replaced an older form for the accompaniment of song and dance called flamenco, and a modified version, known as the flamenco guitar, was created.

III. Performing Techniques

The modern classical guitar is usually played in a seated position, with the instrument resting on the left lap—and the left foot placed on a footstool. Alternatively—if a footstool is not used—a guitar support can be placed between the guitar and the left lap (the support usually attaches to the instrument's side with suction cups).

Right-handed players use the fingers of the right hand to pluck the strings, with the *thumb* plucking from the top of a string downwards (downstroke) and the other *fingers* plucking from the bottom of the string upwards (upstroke). The *little finger* is used only to ride along with the ring finger without striking the strings and to thus physiologically facilitate the ring finger's motion.

Direct contact with strings

As other plucked instruments, the musician directly touches the strings (usually plucking) to produce the sound. This has important consequences: Different tone/timbre (of a single note) can be produced by plucking the string in different manners (apoyando or tirando[9]) and in different positions (such as closer and further away from the guitar bridge). For example, plucking an open string will sound brighter than playing the same note(s) on a fretted position (which would have a warmer tone).

Fingering notation

In guitar scores the five fingers of the right-hand (which pluck the strings) are designated by the first letter of their Spanish names namely p = thumb (pulgar), i = index finger (índice), m = middle finger (mayor), a = ring finger (anular), c = little finger or pinky (meñique/chiquito).

The four fingers of the left hand (which fret the strings) are designated 1 = index, 2 = major, 3 = ring finger, 4 = little finger. 0 designates an open string—a string not stopped by a finger and whose full length thus vibrates when plucked. It is rare to use the left hand thumb in performance.

Alternation

To achieve tremolo effects and rapid, fluent scale passages, the player must practice *alternation*—that is, never plucking a string with the same finger twice in a row. Using **p** to indicate the thumb, *i* the index finger, *m* the middle finger and *a* the ring finger, common alternation patterns include:

i-m-i-m: Basic melody line on the treble strings. Has the appearance of "walking along the strings". This is often used for playing scale (music) passages.

p-i-m-a-i-m-a: Arpeggio pattern example. However, there are many arpeggio patterns incorporated into the classical guitar repertoire.

p-a-m-i-p-a-m-i : Classical guitar tremolo pattern.

p-m-p-m : A way of playing a melody line on the lower strings.

IV. Musical Classics

Memories of the Alhambra   Francisco Terega
　　　　　　　　　　　（《阿尔罕布拉宫的回忆》 弗朗西斯科·泰雷加）
Arabian Style Fantasy   Francisco Terega
　　　　　　　　　　　（《阿拉伯风格绮想曲》弗朗西斯科·泰雷加）
Tears   Francisco Terega　　　　　　　（《泪》弗朗西斯科·泰雷加）
Adilida   Francisco Terega　　　　　　（《阿狄利达》弗朗西斯科·泰雷加）
Morning Song   Francisco Terega　　　（《晨之歌》弗朗西斯科·泰雷加）
Gran Jota de concierto   Francisco Terega   （《大霍塔舞曲》弗朗西斯科·泰雷加）
Magic Flute Variations   Gernando Sor   （《魔笛变奏曲》 费尔南多·索尔）
Moonlight   Gernando Sor   　　　　　（《月光》费尔南多·索尔）
Great Solo   Gernando Sor   　　　　　（《伟大的独奏》费尔南多·索尔）
The Oath of America   Mikael Lyubet   （《阿美利亚的誓言》米凯尔·柳贝特）
The Dance of the Peacock   Luis Millan   （《孔雀之舞》路易斯·米兰）
"Watch the Bull" Variations   Luis de Navais
　　　　　　　　　　　（《"看牛吧"变奏曲》路易斯·德·纳瓦埃斯）
Paganini's Sonata   Niccolò Paganini
　　　　　　　　　　　（《帕格尼尼的的奏鸣曲》尼科罗·帕格尼尼）

📖 Notes on the text

1. acoustic guitar 民谣吉他
2. flamenco guitar 弗拉门戈吉他
3. Antonio Torres Jurado 安东尼奥·托雷斯·朱拉多

4. rasgueado (also called Rageo, spelled so or Rajeo) 延续弹奏
5. vihuela 比尤埃拉琴；比维拉琴（西班牙文艺复兴时期的第一种弹拨乐器）
6. Alonso Mudarra 阿隆索·穆达拉 (1510—1580), 西班牙作曲家。
7. five-string 五根单弦
8. Antonio de Torres Jurado 安东尼奥·迪·托雷斯 (1817—1892), 西班牙的吉他制作家。
9. apoyando (or tirando) 靠弦演法；平行弹法

🎻 Terms

peghead	/ˈpeghed/	琴头
tuning machine	/ˈtuːnɪŋ məˈʃiːn/	弦轴
nut	/nʌt/	上弦枕
fretboard/ fingerboard	/ˈfretbɔːrd/, /ˈfɪŋgərbɔːrd/	指板
fret	/fret/	品丝；品
neck	/nek/	琴颈
heel	/hiːl/	琴肩
soundboard	/saʊnd bɔːrd/	面板
rosette	/rəʊˈzet/	音孔环饰
sound hole	/saʊnd həʊl/	音孔
saddle (bridge nut)	/ˈsædl/	下弦枕
bridge	/brɪdʒ/	琴马；琴桥
body side	/ˈbɑːdi saɪd/	侧板
bottom deck	/ˈbɑːtəm dek/	背板
pluck	/plʌk/	弹拨
downstroke	/ˈdaʊnˌstroʊk/	拨
upstroke	/ˈʌpˌstroʊk/	勾
scale	/skeɪl/	音阶
arpeggio	/ɑːrˈpedʒiəʊ/	琶音
tremolo	/ˈtremələʊ/	轮指

Exercises

I. Comprehension questions

1. What is the classical guitar also known as?
2. How many members does a guitar family mainly include?
3. How many strings does a classical guitar have?
4. Who gave the modern classical guitar its definitive form?
5. What are the functions of the bridge in the classical guitar?

II. Translating useful expressions

1. 古典吉他是拨弦乐器，有六根弦。
2. 古典吉他可用于演奏古典乐曲和其他风格的音乐作品。
3. 古典吉他的六根弦都是由尼龙或尼龙包裹金属制作。
4. Each fret of the classical guitar produces a different pitch and each pitch spaced a half-step apart on the 12 tone scale. Every twelve frets represent one octave.
5. In the middle of the 16th century, influences from the vihuela and the Renaissance guitar were combined and the five-course double-string baroque guitar appeared in Spain.
6. Right-handed players use the fingers of the right hand to pluck the strings, with the thumb plucking from the top of a string downwards (downstroke) and the other fingers plucking from the bottom of the string upwards (upstroke).

III. Brainstorm

The classical guitar and *ruan*, a Chinese indigenous musical instrument, are both plucked-string instruments. Please research and compare the common and different points between the classical guitar and *ruan* from their development, constructions and playing techniques.

Lesson Seven
Brass-Wind Instruments

Lesson Seven
Brass-Wind Instruments

Trumpet

The trumpet, a brass instrument, is an extremely old instrument in the brass family. The English word "trumpet" from Old French "*trompette*," meaning "triumph", was first used in the late 14th century. A trumpet produces a "buzzing" sound by lip vibration and its music is written in treble clef. The trumpet group ranges from the piccolo trumpet with the highest register in the brass family to the bass trumpet which is pitched one octave below the standard B♭ or C trumpet. The trumpet can not only play strong, sharp and brilliant tones, but also plays beautiful melodies, so it is commonly used in classical and jazz ensembles.

I. Construction

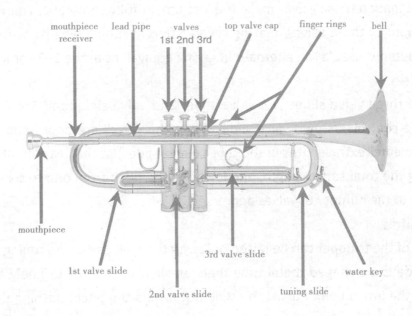

The trumpet is constructed of brass tubing, usually bent twice into a rounded rectangular (圆弧矩形) shape. The trumpet mainly consists of three parts: mouthpiece, pipes and mechanical parts. The tubing have a length of about 1.48 m (4 ft 10 in) (B♭ trumpet).

Mouthpiece

The mouthpiece which is a removable part has a circular rim, which provides a comfortable environment for the lips' vibration. Directly behind the rim is a deep, almost v-shaped cup, which channels the air into a much smaller opening (the back bore or shank) that tapers out slightly to match the diameter of the trumpet's lead pipe. The dimensions of these parts of the mouthpiece affect the timbre or quality of sound, the ease of playability, and player comfort. Generally speaking, the wider and deeper the cup is, the darker the sound and timbre become.

Valve

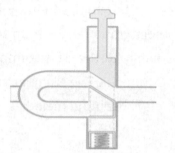

trumpet valve bypass (depressed)

Modern trumpets have three (or, infrequently, four) piston valves, each of which increases the length of tubing when engaged, thereby lowering the pitch. The first valve lowers the instrument's pitch by a whole step (two semitones), the second valve by a half step (one semitone), and the third valve by one and a half steps (three semitones). When a fourth valve is present, as some piccolo trumpets, it usually lowers the pitch a perfect fourth (five semitones). Used singly or in combination these valves make the instrument fully chromatic. There are eight combinations of three valves, making seven different tubing lengths, with the third valve sometimes used as an alternate fingering equivalent to the 1–2 combination.

Valve slide

There are three valve slides: 1st valve slide, 2nd valve slide, and 3rd valve slide. Each valve controls its own valve slide. When the valves are depressed, the air is allowed to enter extra lengths of the tubes (valve slides) of the instruments, thereby increasing the total sound length, and lowering the pitch from one to six semitones according to the number of valves depressed.

Tuning slide

The pitch of the trumpet can be raised or lowered by the use of the tuning slide. The tuning slide is a c-shaped metal tube that can slide in and out to finely adjust the tuning of the instrument. Pulling the slide out lowers the pitch; pushing the slide in raises it.

Lead pipe

The lead pipe is the trumpet part that the player attaches the mouthpiece to.

Finger ring (little finger hook)

The finger ring helps the player to support the trumpet throughout playing. This makes a more comfortable playing, and enables the other hand of the player to be free to make adjustments or turn the pages of sheet music.

Bell

The bell is where the sound of the instrument comes out; it acts as a speaker and projects the sound. Alterations to the bell and the size of the bell will affect its sound. Smaller bell flares sound sharper while bigger flares sound mellower.

Water key

The water key is on the end of the tuning slide. The water key is a simple valve or tap, depending on the instrument, and it allows any build up fluid and saliva to drain out of the instrument.

II. Types

In the trumpet family, the most common trumpet is B♭(a transposing instrument), but A, C, D, E♭, E, low F, and G trumpets are also available. There are two types of valves in the modern trumpets: the piston-valved type[1] and the rotary-valved type[2], and the piston -valved trumpet is more common.

III. Development

The origins of the trumpet are unclear, but the earliest trumpet-like instruments which were mainly made from natural materials, such as conch shells and animal horns, could date back to 1500 BCE and earlier. The Shofar[3], made from a ram horn and the Hatzotzeroth, made of metal, are both mentioned in the Bible. They were played in Solomon's[4] Temple around 3000 years ago and they were said to be used to blow down the walls of Jericho[5]. Some ancient metal (e.g. bronze, silver etc.) trumpets have also been found in Egypt, Scandinavia, Central Asia and China. In ancient times, the trumpets were typically used for signaling in battle and hunting, or religious and military purposes rather than music in the modern sense, and the modern bugle continues this signaling tradition.

Improvements in metal making and instrument design made the trumpet a musical instrument during the late medieval times and the Renaissance. The natural trumpet[6] of these eras had primarily been constructed of a single coiled tube without valves. Changing keys required the player to change crooks[7] of the instruments. The natural trumpet popularized in baroque era which was known as the "Golden Age of the natural trumpet". During this period, a lot of music was written for virtuoso trumpeters. Without pitch-altering devices and valves, all of these trumpets could only produce the notes of a single overtone series. The limitations of these trumpets led to its decline of the popularity during the Classical and Romantic Eras.

In the mid-19th century, the introductions of valves and keys (按键) brought new life to the trumpet, leading to its frequent use in orchestra works. But it was not until 1862, when the Englishman Andir improved the valves and key, that the modern trumpet with three (or, infrequently, four) valves came into being. By the introduction of "valves", this trumpet can play all the twelve semitones in all octaves and become melody-playing instrument. Now, the trumpet has been a core component of the orchestra and bands, and is favored by the people all over the world.

IV. Performing Techniques

How does the trumpet produce the sound?

As with all brass instruments, by blowing air through nearly-closed lips into the mouthpiece, the trumpet produces a "buzzing" sound and starting a standing wave vibration in the air column inside the instrument. The player can select the pitch from a range of overtones (or harmonics) by changing the lip aperture and tension (known as the embouchure).

Finger

The fingering schema arises from the length of each valve's tubing (a longer tube produces a lower pitch). Valve "1" increases the tubing length enough to lower the pitch by *one whole step* (a note), valve "2" by *one half step* (a semitone), and valve "3" by one and a half steps. For example, as showed in the picture, "open" means all valves up, "1" means first valve, "1–2" means first and second valve simultaneously, and so on.

Mute

Being placed in or over the bell, various types of mutes of the trumpet can decrease volume and create mysterious timbre. Of all brass instruments, the trumpets have the widest selection of mutes: the straight mute, cup mute, harmon mute, pluger, bucket mute and wah-wah mute, and the most common type is a straight mute.

Main performing techniques:

Flutter tonguing: The trumpeter rolls the tip of the tongue to produce a "growling like" tone.

Single tonguing: The player articulates using the syllable ta ta ta.

Double tonguing: The player articulates using the syllables ta-ka ta-ka ta-ka.

Triple tonguing: The same as double tonguing, but with the syllables ta-ta-ka ta-

ta-ka ta-ta-ka or ta-ka-ta ta-ka-ta.

Glissando: Trumpeters can slide between notes by depressing the valves halfway and changing the lip tension. Modern repertoire makes extensive use of this technique.

The other important playing techniques have **vibrato, trill** and **circular breathing**.

On trumpet, cornet and flugehorn

The family of the treble brass instruments mainly has piccolo trumpet, bass trumpet, slide trumpet, pocket trumpet, herald trumpet, cornet and flugehorn etc. The trumpet is often confused with its close relative the cornet, which has a more conical tubing shape compared to the trumpet's more cylindrical tube. This, along with additional bends in the cornet's tubing, gives the cornet a slightly mellower tone. The trumpet and the cornet are nearly identical because they have the same length of tubing and the same pitch, and the music written for cornet and trumpet is interchangeable. Another relative, the flugelhorn, has tubing that is even more conical than that of the cornet, and an even richer tone. It is sometimes augmented with a fourth valve to improve the intonation of some lower notes.

V. Musical Classics

Trumpet Concerto in E-flat Major Franz Joseph Haydn
　　　　　　　　　（《降E大调小号协奏曲》弗朗茨·约瑟夫·海顿）
The Turkish March Wolfgang Amadeus Mozart
　　　　　　　　　（《土耳其进行曲》沃尔夫冈·阿玛多伊斯·莫扎特）
Carmen Overture Georges Bizet　　　　　　　（《卡门序曲》乔治·比才）
Trumpet Concerto in E-flat Major Johann Nepomnk Hummel
　　　　　　　　　（《降E大调小号协奏曲》约翰·尼波默克·胡梅尔）
Spanish Matador March Pascual Marquina Narro
　　　　　　　　　（《西班牙斗牛士进行曲》帕斯夸尔·玛奎纳·纳罗）
Red River Valley a Canadian folk song　　　　（《红河谷》加拿大民歌）
From the New World Antonín Leopold Dvořák
　　　　　　　　　（《自新大陆》安东·利奥波德·德沃夏克）
Neapolitan Dance Peter Ilyich Tchaikovsky
　　　　　　　　　（《拿波里舞曲》彼得·伊里奇·柴可夫斯基）
Sabre Dance Aram Ilitch Khatchaturian
　　　　　　　　　（《马刀舞曲》阿拉姆·伊里奇·哈恰图良）

Notes on the text

1. the piston-valved type 直升式活塞
2. the rotary-valved type 回旋式活塞
3. Shofar 羊角号（犹太人用于宗教仪式和古代作战的信号）。
4. Solomon 所罗门，古代以色列国王，出生于公元前1000年，于公元前931年去世。他是大卫之子和继承人，约公元前960—前930年在位。
5. The walls of Jericho 耶利哥之墙，约旦古城，引用于"耶利哥城墙之坚，不足以抵挡其音"。
6. natural trumpet 自然小号，小号的前身，是一种不带阀键的铜管乐器，杯形吹嘴，它只能吹奏泛音，无法演奏半音阶。
7. crook（铜管乐器中可拆下的）变音插管；定调管

Terms

mouthpiece	/ˈmaʊθpiːs/	号嘴
lead pipe	/liːd paɪp/	号嘴导管
bell	/bel/	喇叭口
water key	/ˈwɔːtər kiː/	水门
tuning slide	/ˈtuːnɪŋ slaɪd/	调音管
valve	/vælv/	活塞
valve slide	/vælv slaɪd/	活塞管
finger ring	/ˈfɪŋɡər rɪŋ/	指环；指钩
valve cap	/vælv kæp/	活塞帽
mute	/mjuːt/	弱音器
crook	/krʊk/	变音插管；定调管
flutter tonguing	/ˈflʌtər ˈtʌŋɪŋ/	舌颤音；花舌音
single tonguing	/ˈsɪŋɡl ˈtʌŋɪŋ/	单吐
double tonguing	/ˈdʌbl ˈtʌŋɪŋ/	双吐
triple tonguing	/ˈtrɪpl ˈtʌŋɪŋ/	三吐
glissando	/ɡlɪˈsændəʊ/	滑音
vibrato	/vɪˈbrɑːtəʊ/	（唇）颤音
trill	/trɪl/	（指）颤音
circular breathing	/ˈsɜːrkjələr ˈbriːðɪŋ/	循环呼吸

Exercises

I. Comprehension questions

1. Which type of trumpet has the highest register in the brass family?
2. What was the trumpet typically used for in ancient times? When did the trumpet become a musical instrument?
3. How do the natural trumpets change keys?
4. What are the two types of valves in the modern trumpets? Which one is more common?
5. The modern trumpets usually have three piston valves, what are their functions?

II. Translating useful expressions

1. 小号属铜管乐器，在高音谱号记谱。在小号家族中，降 B 调小号最常用。
2. 小号号嘴各部位的尺寸对音色、音质、演奏的容易度和演奏者的舒适度都有影响。
3. 弱音器可以降低小号的音量，产生神秘的音色。
4. Each valve, when engaged, increases the length of tubing, lowering the pitch of the instrument.
5. The modern trumpets with valves can play all the twelve semitones in all octaves and become melody-playing instrument.
6. By blowing air through nearly-closed lips into the mouthpiece, the trumpet produces a "buzzing" sound and starting a standing wave vibration in the air column inside the instrument.

III. Brainstorm

Without pitch-altering devices and valves, the natural trumpets could only produce the notes of a single overtone series, so the limitations of these trumpets lead to its decline of the popularity during the Classical and Romantic Eras. Until 1862, the Englishman Andir brought new life to the natural trumpet by inventing valves and keys. **Development is a top priority**. In combination with your own experiences, give a specific example to brainstorm the importance of development to everyone, even every nation.

Horn

Horn and French horn

In a broad sense, the horn is a family of musical instruments made of a tube, which is usually made of metal and often curved in various ways, with one narrow end into which the musician blows, and a wide end from which sound emerges. The bore of the horn is conical rather than cylindrical. The variety in horn history includes fingerhole horns, the natural horn, Russian horns, French horn, German horn, Vienna horn, mellophone, marching horn, and Wagner tuba etc. In jazz and popular-music contexts, the word "horn" may be used loosely to refer to any wind instrument, and a section of brass or woodwind instruments, or a mixture of the two.

The name "French horn" first came into use in the late 17th century. At that time, French makers were preeminent in the manufacture of hunting horns, and they were considered to invent the now-familiar circular "hoop" shape of the instrument, so the French horn is referred simply as the "horn" in professional music circles since the 1930s. The International Horn Society[1] has recommended that the French horn be simply called the horn since 1971.

The French horn (or simply **horn**) is a lip vibration instrument and a transposing instrument. The horn is a circular valved brass instrument which is made of tubing wrapped into a coil with a flared bell, a conical bore, a funnel-shaped mouthpiece. There are two special places about the horn: firstly it's played with the left hand and the right hand, and the right hand can be used to change the notes; secondly the conical bore, which starts small and winds around and gradually increases in width to the bell, makes the horn slightly different from other brass instruments. Because of its extremely rich expression, the horn has been the link between brass and woodwind instruments in ensemble. Because of its widest range, the horn is called "the king of the orchestra".

I. Construction

A horn mainly consists of three parts: mouthpiece, pipes and mechanical parts. Most horns have lever-operated rotary valves, but some, especially older horns, use piston valves. The tubing in F horn is 3.930meters long, and it is curved into a circular.

1. mouthpiece 号嘴

2. leadpipe 导管 : where the mouthpiece is placed
3. first rotary valve 第一旋转式活塞
4. main slide 主管
5. F slide F 调滑管
6. second slide /first slide 第二滑管 / 第一滑管
7. bell 喇叭口
8. second rotary valve/third rotary valve/fourth rotary valve 第二旋转式活塞 / 第三旋转式活塞 / 第四旋转式活塞 : operated with the left hand
9. third lever/second lever/first lever 第一按键 / 第二按键 / 第三按键
10. fourth lever 转调键 : to switch between F and B♭

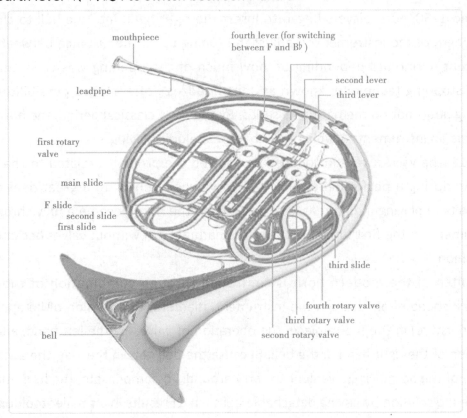

II. Development

The horn originated in Europe and its ancestry can be traced back more than 3,000 years. The earliest horn was made of the actual horns of animals rather than metal or other materials. According to the historical records, the shofar, a ram's horn, is the origin of the horn, and it plays an important role in Jewish religious rituals.

Early metal horns were less complex than modern horns. The modern horn was developed from the French hunting horn around 1650. To allow the instruments to be played on horseback in hunting, the hunting horns were made in a tightly coiled spiral form, and the mouthpiece was irremovable by the early 17th century. These early hunting horns without valves, which were called natural horns, could only play the harmonic series. The remedy for the limitations was the use of crooks, i.e., tubing of differing length that, when inserted, altered the length of the instrument, and thus its pitch. By the second decade of the 18th century, because of its poetic and satisfying sound, the natural horns became very popular in orchestra, church music and chamber music.

Around 1750, horn players began to insert the right hand into the bell to change the length of the instrument to adjust the tuning up to the distance between two adjacent harmonics depending on how much of the opening was covered. This hand-stopping technique, known as sons boudzes[2], offers more possibilities for playing notes not on the harmonic series. By the early classical period, the horn had become an instrument capable of much more melodic playing.

In 1818, the valves were introduced and solved the problems related to changing crooks during a performance. The use of valves opened up a great deal more flexibility in playing in different keys, and makes the horn become a fully chromatic instrument for the first time. Gradually, the natural horn without valves became out of fashion.

The pitch of the modern horn is controlled through combination of the four factors: speed of air through the instrument, diameter and tension of lip aperture (embouchure) in the mouthpiece, the operation of valves by the left hand, and the position of the right hand in the bell. Because the bell gets in the way, the awkward shape of the horn is inconvenient to carry around. To compensate, the horn makers create the solution by having detachable bells which results in a smaller toolbox.

III. Types

The horns may be mainly classified in single horn, double horn and triple horn, but the single horn and the double horn are more common.

Single horns use a single set of tubes connected to the valves. The three valves control the flow of air in the single horn, which is tuned to F or less commonly B♭.

F and B♭ horns are simple in use, light in weight, and cheap in price, so they are still used today as student models. The accuracy of F horns has some problems in the highest range while the B♭ horns have a less desirable sound in the mid and especially the low register.

Double horn can solve the problems produced by the single horns. In 1897, double horn was invented by the German horn maker Ed. Kruspe. The double horn combines two instruments into a single frame: the original horn in F, and a second, higher horn keyed in B♭. The more common double horn has a fourth, trigger valve, usually operated by the thumb. By using a fourth valve, the horn player can quickly switch from the deep, warm tones of the F horn to the higher, brighter tones of the B♭ horn, or vice versa.

IV. Performing Techniques

An important element in playing the horn is to deal with the mouthpiece. Most of the time, the mouthpiece is placed in the exact center of the lips, but, because of differences in the formation of the lips and teeth of different players, some tend to play with the mouthpiece slightly off center, e.g. the exact side-to-side placement of the mouthpiece, the up-and-down placement of the mouthpiece.

There are several special playing techniques about the horn, which include placing the mutes (con sord)[3] in the bell, putting the right hand into the bell to play the sons boudzes, and pressing the lips against the mouthpiece to play the cuivre[4]. Compared with the sons boudzes, when playing the cuivre, the player should insert the right hand deeper into the bell.

The other important performing techniques are breathing, single tonguing, double tonguing, triple tonguing, trill, vibrato, legato, and staccato.

V. Musical Classics

Horn Concerto No. 1, 2, 3, and 4 Wolfgang Amadeus Mozart
　　　　（《第一、二、三、四圆号协奏曲》沃尔夫冈·阿玛多伊斯·莫扎特）
Horn Concertos Richard Georg Strauss　　（《圆号协奏曲》理查德·施特劳斯）
The Fire Bird (Suite) Igor Strvinsky　　　（《火鸟》组曲伊戈尔斯特拉文斯基）

Notes on the text

1. The International Horn Society 国际圆号协会
2. sons boudzes 阻塞音
3. mutes (con sord) 弱音器
4. cuivre 闭塞音

Terms

mouthpiece	/ˈmaʊθpiːs/	号嘴
leadpipe	/ˈliːdpaɪp/	导管
rotary valve	/ˈrəʊtəri vælv/	转式活塞
main slide	/meɪn slaɪd/	主管
bell	/bel/	喇叭口
fourth lever	/fɔːrθ ˈlevər/	转调键
crook	/krʊk/	变音插管；定调管
single tonguing	/ˈsɪŋɡl ˈtʌŋɪŋ/	单吐音
double tonguing	/ˈdʌbl ˈtʌŋɪŋ/	双吐
triple tonguing	/ˈtrɪpl ˈtʌŋɪŋ/	三吐
legato	/lɪˈɡɑːtəʊ/	连奏
staccato	/stəˈkɑːtoʊ/	断奏
vibrato	/vɪˈbrɑːtəʊ/	（唇）颤音
trill	/trɪl/	（指）颤音

Exercises

I. Comprehension questions

1. Why is the French horn referred to simply as the "horn" in professional music circles since the 1930s?
2. What are the two special places about the horn?
3. What is the role of the horn in an orchestra?
4. How does the crook function?
5. What is the difference between the natural horn and the modern one?

II. Translating useful expressions

1. 圆号是唇振乐器和移调乐器。由于圆号丰富的表现力，它在合奏中能与木管乐器和弦乐器的声音很好的融合。
2. 圆号演奏者可以把右手插进圆号的喇叭口来改变其音高。
3. 圆号的三个阀键单独或合并使用 (singly or in combination)，可吹奏出音域内所有半音阶。
4. The natural horns without valves could only play the harmonic series and have no ability to play in different keys, so the keys of a natural horn could be changed by adding different crooks of different length.
5. The pitch of the modern horn is controlled through combination of the four factors: speed of air through the instrument; diameter and tension of lip aperture in the mouthpiece; the operation of valves by the left hand, and the position of the right hand in the bell.
6. An important element in playing the horn is to deal with the mouthpiece. Three main ways of the mouthpiece are placed: in the exact center of the lips, the exact side-to-side placement, and the up-and-down placement.

III. Brainstorm

Horn players began to insert the right hand into the bell to change the pitch around 1750. This *special* playing technique offers more possibilities for the horn, so it has become an instrument capable of much more melodic playing by the early classical period. From this *special* point on the horn, tell us a *special* story about yourself, on how the *special* experience has inspired you to grow up and become better yourself.

Trombone

The trombone, also called slide trombone, is a lip vibration instrument belonging to the brass family. Unlike most other brass instruments, which have valves that, when pressed, alter the pitch of the instrument, trombones, as the only brass instrument with few changes in construction, have a telescoping slide mechanism that varies the length of the instrument to change the pitch. However, many modern trombone models also have a valve attachment which lowers the pitch of the instrument. The trombones have used slides since their inception, so they have always been fully chromatic instruments. Because of its unique timbre, the trombone plays an important role in orchestras, military bands, and brass bands, and it is widely used in jazz bands, which is known as the "king of jazz".

I. Construction

The trombone is a predominantly cylindrical tube bent into an elongated "S" shape. Rather than being completely cylindrical from end to end, the tube is a complex series of tapers with the smallest at the mouthpiece receiver and the largest just before the bell flare. The design of these tapers affects the intonation of the instrument.

A trombone mainly consists of mouthpiece, slides and bell, and there are some variations in construction.

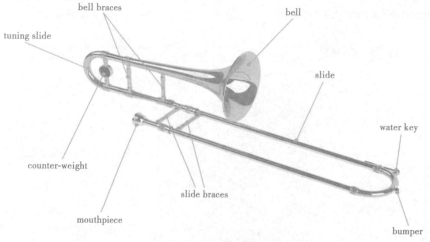

trombone without F-valve

Lesson Seven
Brass-Wind Instruments

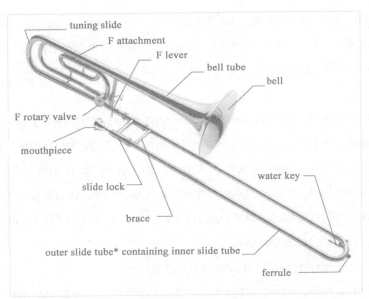

trombone with F-valve

Mouthpiece

The detachable cup-shaped mouthpiece is inserted into the mouthpiece receiver in the slide section. Variations in mouthpiece construction affect the individual player's ability to make a lip seal and produce a reliable tone, the timbre of that tone, its volume, the instrument's intonation tendencies, the player's subjective level of comfort, and the instrument's playability in a given pitch range.

Slide

The slide, the most distinctive feature of the trombone, allows the player to extend the length of the air column, lowering the pitch. The slide section consists of a leadpipe, the inner and outer slide tubes, and the bracing, or "stay". To prevent friction from slowing the action of the slide, additional sleeves known as "stockings" were developed. These "stockings" were soldered onto the ends of the inner slide tubes. Nowadays, the stockings are incorporated into the manufacturing process of the inner slide tubes and represent a fractional widening of the tube to accommodate the necessary method of alleviating friction. This part of the slide must be lubricated frequently.

Additional tubing connects the slide to the bell of the instrument through a neckpipe, and bell or back bow (U-bend). The joint connecting the slide and bell sections is furnished with a *threaded collar* (螺纹环) to secure the connection of the two parts of the instrument.

The adjustment of intonation is most often accomplished with a short tuning slide between the neckpipe and the bell. However, trombonists, unlike other instrumentalists, are not subject to the intonation issues resulting from valved or keyed instruments, since they can adjust intonation "on the fly" by subtly altering slide positions when necessary.

Bell

Trombone bells may be constructed of different brass mixtures. The most common tenor trombone bells are usually from 19 cm to 22 cm in diameter, while bass trombone bells are usually either 24 cm or 25 cm in diameter. The bell may be constructed out of two separate brass sheets or out of one single piece of metal, and the edge of the bell may be finished with or without a piece of bell wire to secure it, which also affects the tone quality.

Variations in construction: Valve attachments

Many trombones have valve attachments to aid in increasing the range of the instrument while also allowing alternate slide positions for difficult music passages. In addition, valve attachments make trills much easier. Valve attachments appear on alto, tenor, bass, and contrabass trombones. The most common type of valve seen for valve attachments is the rotary valve, and the piston valve is very rare.

II. Types

The **trombone** (also called **slide trombone**) has been built in sizes from piccolo to contrabass. There are mainly four types of trombones: alto trombone (E-flat), tenor trombone (B-flat without F-valve[1]), tenor trombone (B-flat with F-valve[2]), and bass trombone (B-flat with F-valve and G-valve, and a larger bell[3]). The most common trombones are the tenor trombone and bass trombone.

Some slide trombones have one or (less frequently) two rotary valves operated by a left-hand thumb trigger. The single rotary valve is part of the F attachment, which adds a length of tubing to lower the instrument's fundamental pitch from B♭ to F. Some bass trombones have a second trigger with a different length of tubing. The second trigger facilitates playing the otherwise problematic low B.

Slide trombones are usually constructed with a slide that is used to change the pitch. **Valve trombones** use three valves (singly or in combination) instead of the slide. The valves follow the same schema as other valved instruments—the first valve lowers the pitch by one step, the second valve by a half-step, and the third valve by one and a half steps.

III. Development

The history of the trombone can date back to the 15th century. About 1700 years ago, the trombone was called the sackbut[4]. The sackbut was very popular in Europe between the 15th century and the mid-late 17th century. As the slide of the sackbut became longer, the instrument was known as the "trombone". The 17th-century trombone was made in slightly smaller dimensions and had a more conical and less flared bell than modern trombone. In the 17th and 18th centuries it was used in the supernatural scenes of sacred works and in operas, for example, both Bach and Handel[5] employed the trombone in their works. The first use of the trombone in a symphony was Ludwig van Beethoven[6] in his *Symphony No. 5 in C Minor*[7].

In 19th century, Sattler improved the design of the trombone. Sattler's reforms to the trombone had the snake decorations, the bell garland, the wide bell flare, and the most significant widening of the bore. The further changes to the trombone in this period were the addition of "stocking" at the end of the inner slide to reduce friction, the development of the water key to expel condensation from the horn, and the occasional addition of a valve that, intentionally, only was to be set on or off but later was to become the regular F-valve. By about the mid-19th century, the trio of two tenor trombone and one bass became standard in orchestra.

In the 20th century the trombone maintained its important place in the orchestras, wind bands and jazz. During the 20th century, improvements in construction were the use of different materials, the increase of dimension in mouthpiece, bore and the bell, and the increase of types of mutes and valves. One of the most crucial changes is the popularity of the F-attachment trigger[8] which can make the trombone more convenience and versatility.

IV. Performing Techniques

The trombone is operated by the right hand of the player, and the sound is produced when the player's vibrating lips (embouchure) causes the air column inside the instrument to vibrate.

Basic slide positions: The modern system has seven chromatic slide positions on a tenor trombone in B-flat. Each successive position outward (approximately 8 cm) will produce a note which is one semitone lower when played in the same partial.

Glissando: By moving the slide without interrupting the airflow or sound production, the trombone can produce a true glissando. A tritone is the largest interval that can be performed as a glissando.

Trill: Though generally simple with valves, trills are difficult on the slide trombone. Trills tend to be easiest and most effective higher in the harmonic series because the distance between notes is much smaller and slide movement is minimal.

Mute: A variety of mutes can be used with the trombone to alter its timbre. Many mutes are held in place with the use of cork grips, including the straight, cup, and harmon. Some fit over the bell, like the bucket mute, and some can be held in front of the bell and moved to cover more or less area for a wah-wah effect.

The other important performing techniques are breathing, single tonguing (单吐), double tonguing.

V. Musical Classics

Romance   Carl Maria von Weber   (《浪漫曲》卡尔·马利亚·冯·韦伯)
Concerto in E Flat Major F. C. David   (《降 E 大调协奏曲》达维德)
Serenade   Franz Schubert   (《小夜曲》弗朗茨·舒伯特)
Seventy Six Trombone   Meredith Willson   (《76 只长号》梅雷迪思·威尔森)
The Flight of the Bumblebee   Nikolai Andreivich Rimsky-Korsakov
        (《野蜂飞舞》尼古拉·安德烈伊维奇·里姆斯基-科萨科夫)
Grondahl   Ma Youdao   (《格隆达尔》马友道)
The Rising Sun on the Grassland   Meiliqige
        (《草原上升起不落的太阳》美丽其格)

🎹 Notes on the text

1. tenor trombone (B-flat without F-valve) 次中音长号（降 B 调不带 F 变音键的次中音长号）
2. tenor trombone (B-flat with F-valve) 次中音长号（降 B 调带 F 变音键的次中音长号）

3. bass trombone (B-flat with F-valve and G-valve, and a larger bell) 低音长号（降B调带F、G两个变音键的低音长号，号的喇叭口比一般的号更大）
4. sackbut 萨克布，是古老的铜管乐器中的一种，源于公元15世纪初期的欧洲。早期的长号不具备现在的活塞装置，早期的长号也叫萨克布，它主要靠内外套管的伸缩来决定演奏的音高。
5. George Friedrich Handel 乔治·弗里德里希·亨德尔（1685—1759），出生于德国哈雷，巴洛克时期英籍德国作曲家。
6. Ludwig van Beethoven 路德维希·凡·贝多芬（1770—1827），出生于波恩，维也纳古典乐派代表人物之一，欧洲古典主义时期作曲家。
7. *Symphony No. 5 in C Minor* 《C小调第五交响曲》
8. F-attachment trigger F键/转调键

Terms

tuning slide	/ˈtuːnɪŋ slaɪd/	调音管
counterweight	/ˈkaʊntərweɪt/	平衡器
mouthpiece	/ˈmaʊθpiːs/	号嘴
slide brace	/slaɪd breɪs/	拉杆支撑
slide lock	/slaɪd lɑːk/	拉管锁环
ferrule/ bumper	/ˈferəl/, /ˈbʌmpər/	橡胶垫
water key	/ˈwɔːtər kiː/	放水阀
slide	/slaɪd/	拉管
outer slide tube	/ˈaʊtər slaɪd tuːb/	外拉管
F lever	/ef ˈlevər/	变音键；转调键
F attachment	/ef əˈtætʃmənt/	（转调键控制的）滑管
neckpipe	/ˈnekpaɪp/	颈管
U-bend	/(j)uː bend/	U形弯管
brace	/breɪs/	拉管支杆
glissando	/glɪˈsændəʊ/	滑音
trill	/trɪl/	颤音
single tonguing	/ˈsɪŋɡl ˈtʌnjɪŋ/	单吐
double tonguing	/ˈdʌbl ˈtʌnjɪŋ/	双吐

Exercises

I. Comprehension questions

1. What was the trombone called about 1700 years ago?
2. Who firstly used the trombone in a symphony?
3. What trombones are the most common used?
4. What is the most distinctive feature of the trombone? What is its function?
5. What is the difference between the tenor trombone and the bass one?

II. Translating useful expressions

1. 长号又称拉管号。长号用一个能改变乐器长度的伸缩滑动装置来改变音高。
2. 长号能演奏独特的滑音。
3. 在管弦乐队中常用三支长号，其中两支次中音长号和一只低音长号。
4. In the 20th century, one of the most crucial changes is the popularity of the F-attachment trigger which can make the trombone more convenience and versatility.
5. By moving the slide without interrupting the airflow or sound production, the trombone can produce a true glissando.
6. The modern system has seven chromatic slide positions on a tenor trombone in B-flat. Each successive position outward (approximately 8 cm) will produce a note which is one semitone lower when played in the same partial.

III. Brainstorm

Besides the Western trombone, there are a variety of Chinese national trombones with a long history. Please do some researches on Chinese national trombones, and choose one to state its development, construction and playing techniques in class.

Lesson Seven
Brass-Wind Instruments

Tuba

The tuba is the largest and lowest-pitched musical instrument in the brass family, so most music for tuba is written in bass clef. As with all brass instruments, the sound is produced by lip vibration, or a buzz, into a large mouthpiece. The tuba is a newly-appeared instrument in the mid-19th century, and is one of the newer instruments in the modern orchestra and concert band. The tuba is used mainly as low-pitched harmony and melody and rarely used as a solo instrument in a band. In fact, the sound of the tuba is not only rich in low register, but also beautiful in high register. As the lowest-pitched musical instrument in the brass instruments, it is the solid foundation of the band.

I. Construction

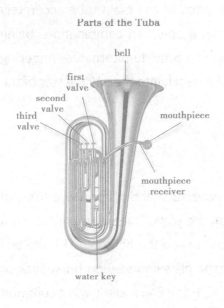
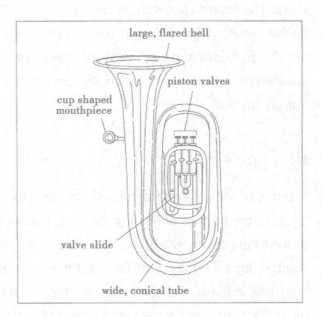

A tuba has four parts: vertically coiled tubing, three to six valves, a flared bell, and a cup-shaped mouthpiece.

Mouthpiece
The mouthpiece is usually located near the bell of the tuba, and it is attached to the small end of the tube.

Tube
As the largest part of the tuba, ranging in length from 12 to 18 feet (3.7 to 5.5 meters),

the tube is considerably long in order to give the instrument its low pitch. There are two shapes of tubes: concert tubas are coiled in an oblong shape, while marching tubas into a circle.

Bell

The conical bell, where the sound comes out, is attached to the tube at the end. The bells of tuba may point directly upward as a concert tuba, pointing backwards over the player's shoulder as marching tuba.

Valve

The tuba has two types of valves: piston valves and rotary valves. The valves add tubing to the main tube of the instrument, thus lowering its fundamental pitch. The tuba usually has three to six valves, and three-valve tubas are designed for amateurs, while four and six for professionals. The first valve lowers the pitch by a whole step (two semitones), the second valve by a semitone, and the third valve by three semitones. Used in combination, the valve tubing is too short and the resulting pitch tends to be sharp. The fourth valve can be tuned to lower the pitch of the main tube accurately by five semitones, and thus its use corrects the main problem of combinations being too sharp. A fifth and sixth valve, if fitted, are used to provide alternative fingering possibilities to improve intonation, and are also used to reach into the low register of the instrument.

II. Types

In terms of the valve of the tuba, there are rotary valve tuba and piston valve tuba. In terms of the posture of holding the tuba, placing on the player's lap is usually called a concert tuba or simply a tuba[1]. The modern sousaphone and the helicon[2], with the bell pointed up, are wrapped to surround the body of the player to play on horseback or marching. In terms of tonality, there are tubas in B♭, C, F and E♭, and the most common tubas is B♭. B♭ and C tubas are mainly used in bands, while F and E♭ tubas in solo. The lowest-pitched tubas are the contrabass tubas.

III. Development

The ophicleide[3] and the serpent[4] were the forerunners of the tuba, which depended on the tone holes to change the pitch. Both of them were limited to notes in the harmonic series, so their sound couldn't match the sound of other brass instruments. Based on the two instruments, after many explorations and improvements by many musicians and instrument-makers, some bass instruments, such as the Bombardon[5], Euphonium[6] and Wagner Tuba[7], appeared successively in the 19th century. In 1835, Wilhelm Friedrich Wieprecht[8] and Johann Gottfried Moritz[9] invented a "bass tuba" with five valves. In 1838, the first tenor tuba was invented by Carl Wilhelm Moritz[10], son of Johann Gottfried Moritz. By using valves to adjust the length of the bugle, the tuba produced a smoother tone that eventually led to its popularity.

With the advancement of technology and the development of music, the different kinds of tubas have been improved and perfected. Adolphe Sax[11] contributed a lot to the improvement of the tuba. Adolphe Sax successfully worked out the dimensional proportion[12] of brass instrument and developed a series of brass instruments (from soprano to bass) known as saxhorns[13], which he patented in 1845. In the late 19th century, the modern sousaphone[14], created by American bandmaster John Philip Sousa[15], has an unique circular shape which makes it very comfortable to carry around on the shoulder, so it becomes popular to play on horseback and the marching band.

IV. Performing Techniques

The playing technique of the tuba is similar to that of other brass instruments, so it is important for players to develop breath. Tuba is a big instrument, so the air needs to be big and fast to get the sound out of the horn. Breathe deeply down into the diaphragm, not up high in the throat. Tense the abdominal muscles while play and blow. That air has a long way to go, so start it out from a place of power.

While blowing, close the lip to the point that it vibrates in the mouthpiece. Keep blowing and vibrating lips so that sound gets out of the tuba. As tuba is a large brass instrument, try blowing a raspberry into the mouthpiece. Main playing techniques have legato, single tonguing, double tonguing, and triple tonguing.

V. Musical Classics

Requiem, K. 626: Tuba Mirum　Wolfgang Amadeus Mozart
　　　　　　　　　　　　　　　　（《安魂曲》沃尔夫冈·阿玛多伊斯·莫扎特）

Czardas　Paul Maurier
　　　　　　　　　　　　　　　　（《查尔达斯舞曲》保罗·莫里埃）

Concerto in F Minor for Bass Tuba and Orchestra   Ralph Vaughan Williams
（《f小调低音大号协奏曲》拉尔夫·沃恩-威廉斯）
Gabriel's Oboe   Ennio Morricone   （《嘉比尔的双簧管》恩尼奥·莫瑞康尼）

Notes on the text

1. concert tuba (or simply tuba) 抱式大号，抱贝司
2. helicon 低音大圆号，海利空低音大号（军乐队等中可套在肩上吹奏的大喇叭）
3. ophicleide 奥菲克莱德号（一种低音金管乐器 / 开孔大号）
4. serpent 蛇形号
5. Bombardon 邦巴东号，低音大号
6. Euphonium 尤风宁号，次中音号
7. Wagner Tuba 瓦纳大号，次中音号
8. Wilhelm Friedrich Wieprecht 维尔海姆·旨里德里赫·维帕列赫特（1802—1872），当时普鲁士的一个军乐队的音乐监督。
9. Johann Gottfried Moritz 约翰·弋特弗里德·莫里茨（1777—1840），德国乐器制造师。
10. Carl Wilhelm Moritz 卡尔·威廉·莫里茨（1810—1855）
11. Adolphe Sax 阿道夫·萨克斯（1814—1894）是萨克斯管这种乐器的发明者。
12. the dimensional proportion 尺寸比例
13. saxhorn 萨克号
14. sousaphone 圈式大号，圈贝司（俗称苏萨风，扛号）
15. John Philip Sousa 约翰·菲力浦·苏萨（1854—1932），美国作曲家、军乐指挥家，对美国铜管乐的发展起了重大的推进作用，被誉为"进行曲之王"。

Terms

mouthpiece	/ˈmaʊθpiːs/	号嘴
tube	/tuːb/	管体
bell	/bel/	喇叭口
valve	/vælv/	活塞
legato	/lɪˈɡɑːtəʊ/	连奏
tonguing	/ˈtʌnɪŋ/	吐音
single/ double/ triple tonguing	/ˈsɪŋɡl ˈtʌnɪŋ/, /ˈdʌbl ˈtʌnɪŋ/, /ˈtrɪpl ˈtʌnɪŋ/	单吐 / 双吐 / 三吐
circular breathing	/ˈsɜːrkjələr ˈbriːðɪŋ/	循环呼吸

Lesson Seven
Brass-Wind Instruments

Exercises

I. Comprehension questions

1. Which is the largest and lowest-pitched musical instrument in the brass family?
2. When was the bass tuba invented?
3. How many valves do the tubas generally have?
4. What is Adolphe Sax's contribution to the musical instruments?
5. How do you describe the modern sousaphone?

II. Translating useful expressions

1. 大号是音最低的铜管乐器，在乐队中主要担当低声部的和声和旋律，很少用于独奏。
2. 从持乐器的姿势上看，大号有抱号和扛号。
3. 大号不仅低音区浑厚，它的高音区也很优美。
4. The tuba has two types of valves: piston valves and rotary valves. The valves add tubing to the main tube of the instrument, thus lowering its fundamental pitch.
5. In terms of tonality, there are tubas in B♭, C, F and E♭, and the most common tubas is B♭. The lowest-pitched tubas are the contrabass tubas.
6. The playing technique of the tuba is similar to that of other brass instruments, so it is important for players to develop breath. Main playing techniques: legato tonguing, single tonguing, double tonguing, and triple tonguing.

III. Brainstorm

If you compare a band to a pyramid, the tuba is the base of the pyramid. The lowest-pitched tuba in the brass family is the solid foundation of the band. If the foundation of the high wall is very low, and even if such a wall has been built, it will surely collapse. (墙高基下，虽得必失。) Please exemplify the importance of solid foundation to everything.

Lesson Eight
Woodwind Instruments

Lesson Eight
Woodwind Instruments

Flute

The flute is a family of musical instruments in the woodwind group. A flute is aerophone or reedless wind instrument that produces its sound when a stream of air directed across a hole in the instrument creates a vibration of air at the hole. Flutes are the earliest known identifiable musical instruments, because a number of flutes dating to about 43, 000 to 35,000 years ago have been found in Europe. Although the earliest flutes were discovered in Europe, in more recent millennia the flute was absent from the continent until it was introduced from Asia about 800 AD.

Types of Flutes

In its most basic form, a flute is an open tube which is blown into. There are several broad classes of flutes. With most flutes (non-fipple flutes), the musician blows directly across the edge of the mouthpiece, with 1/4 of their bottom lip covering the embouchure hole. However, fipple flutes, such as the whistle, flageolet and recorder have a duct that directs the air onto the edge (an arrangement that is termed a "fipple"). The fipple gives the instrument a distinct timbre which is different from non-fipple flutes and makes the instrument easier to play, but takes a degree of control away from the musician.

Another division is between side-blown (or transverse) flutes, such as the Western concert flute, piccolo, dizi and bansuri; and endblown flutes, such as the xiao, Anasazi flute and quena. The player of a side-blown flute uses a hole on the side of the tube to produce a tone, instead of blowing on an end of the tube. End-blown flutes should not be confused with fipple flutes such as the recorder, which are also played vertically but have an internal duct to direct the air flow across the edge of the tone hole.

Flutes may be open at one or both ends. The xun, pan pipes, police whistle, and bosun's whistle are closed-ended. Open-ended flutes such as the concert flute and the recorder have more harmonics, and thus more flexibility for the player, and brighter timbres. An organ pipe may be either open or closed, depending on the sound desired.

Flutes may have any number of pipes or tubes, though one is the most common number. Flutes with multiple resonators may be played one resonator at a time (as is typical with pan pipes) or more than one at a time (as is typical with double flutes). Flutes can be played with several different air sources. Conventional flutes are blown with the mouth, although some cultures use nose flutes.

The flutes in different countries and regions are different, such as Western transverse flutes, Indian flutes, Chinese flutes, Korean flutes, Japanese flutes, Sodina and suling, and Sring. Western transverse flutes mainly include Western concert flute and its variants (e.g. the piccolo).

Western Concert Flute

The Western concert flute is a family of transverse (side-blown) woodwind instruments made of metal or wood, and is the most common variant of the Western transverse flutes. The standard concert flute, also called C flute, Boehm flute, silver flute, or simply flute, is pitched in C and has a range of three octaves starting from middle C or one half step lower when a B foot is attached. Unlike woodwind instruments with reeds, the flute is an aerophone or reedless wind instrument that produces its sound by blowing a stream of air over the embouchure hole. The flute is the main high-pitched melodic instrument in modern orchestral music and chamber music. The tone color of the flute is lively and bright in the high register, and beautiful and pleasant in the low register. The standard concert flute is used in many ensembles, including concert bands, military bands, marching bands, orchestras, and occasionally jazz bands. Other flutes in this family include the piccolo, the alto flute, and the bass flute.

I. Construction

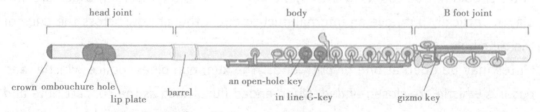

Labelled parts of a French model (open-hole) flute with a B-foot joint.

The flute is composed of three parts: the head joint[1], body, and foot joint.

Head joint
The head joint is the top section of the flute, which has the tone hole and lip plate

where the player initiates the sound by blowing air across the opening. The head joint is sealed by a cork. Gross, temporary adjustments of pitch are made by moving the head joint in and out of the head joint tenon.

Crown: The cap at the end of the head joint that unscrews to expose the cork and helps keep the head joint cork positioned at the proper depth.

Embouchure Hole

Lip Plate (or Embouchure Plate): The part of the head joint that contacts the player's lower lip, allowing positioning and direction of the air stream.

Tenon: The bit of the head joint that goes inside the barrel or head receiver.

Body

The body is the middle section of the flute with the majority of the keys.

Open-hole keys: A finger key with a perforated center

Closed keys (plateau key): A fully covered finger key

Inline G key: The standard position of the left-hand G (third-finger) key which is in line with the first and second keys.

Offset G key: A G key extended to the side of the other two left-hand finger keys, making it easier to reach and cover effectively.

Trill keys: two small, teardrop shaped keys between the right-hand keys on the body.

Split E mechanism: a system whereby the second G key (positioned below the G# key) is closed when the right middle-finger key is depressed, enabling a clearer third octave.

Foot joint

The foot Joint is the last section of the flute (two types: C foot, B foot).

C foot Joint: a foot joint with a lowest note of middle C, which is typically used on student flute.

B foot Joint: a foot joint with a lowest note of B below middle C, which is an option for intermediate and professional flutes.

D# roller: an optional feature added to the Eb key on the foot joint, facilitating the transition between Eb/ D# and Db/C.

II. Development

Origin

The predecessors of the modern concert flute were keyless wooden transverse flutes. About 800 AD, the transverse flute arrived in Europe from Asia via the

Byzantine, where it migrated to Germany and France. These original transverse flutes became known as "German flute" to distinguish them from others, such as the recorder. The flute was used in court music[2] and secular music[3] in the Middle Ages. From Renaissance to the 17th century, the transverse flute, usually made in one section and having a cylindrical bore, began appearing in court, the theatre music, and the first flute solos.

Baroque flute (or traverso)

During the Baroque Era, the transverse flute was redesigned, now often was called the traverso[4]. The wooden one-keyed transverse Baroque flute was made in three or four sections or joints with a conical bore from the head joint down. The conical bore design gave the flute a wide range and more penetrating sound, so it began to be used in opera, ballet, chamber music and orchestral music. The interest in flute increased and peaked in the early half of the 19th century, and the early 19th century saw a great variety of flute designs. Flutes began to lose favor in the Romantic Era, because the symphony orchestra featured brass and strings.

Boehm flute[5]

In the 19th century, a great flutist, composer, acoustician, and silversmith Theobald Boehm[6] began to make flute. Boehm's major innovation to flutes were the change to metal instead of wood, large straight tube bore, the size and placement of tone holes, "parabolic" tapered head joint bore, very large tone holes covered by keys, and the linked key system, which simplified fingering. Boehm's key system remains regarded as the most effective system of any modern woodwind, allowing trained players to perform with facility and extraordinary velocity and brilliance in all keys. With some improvements, Western concert flutes typically conform to Boehm's design, known as the Boehm system, which greatly improved the range and intonation of fllute.

Modified Boehm flute

In the 1950s, Albert Cooper modified the Boehm flute to make playing modern music easier. The flute was tuned to A 440, and the embouchure hole was cut in a new way to change the timbre. In the 1980s, Johan Brogger modified the Boehm flute by using non-rotating shafts, which gave a quieter sound and less friction on moving parts. Also, the modifications allowed for springs to be adjusted individually, and the flute was strengthened.

III. Performing Techniques

Changing pitch

The pitch is changed by opening or closing keys that cover circular tone holes (there are typically 16 tone holes). Opening and closing the holes produces higher and lower pitches. Higher pitches can also be achieved through over-blowing, like most other woodwind instruments. The direction and intensity of the airstream also affects the pitch, timbre, and dynamics.

Breathing

Take a deep breath, and breathe out while slightly smiling, making a "too" sound. Make sure that the shape of your lips doesn't change as you breathe out. Diaphragmatic breathing7 and circular breathing8 are two main breathing techniques for the flute players to use.

Tonguing

Use the head joint, move your tongue as if saying "too-too" while blowing. Pay close attention to motion of your tongue when blowing if you want to produce a good sound. The techniques of tonguing include single tonguing, double tonguing, triple tonguing and flutter tonguing.

Glissando

There are a variety of different glissando techniques. Broadly speaking, glissando techniques can be divided into continuous glissando and discrete glissando, and the latter is more usual for flutist to use.

The other playing techniques have trill, legato and humming at the same time.

IV. Music Classics

Seven Flute Concertos Antonio Vivaldi
（《七首长笛协奏曲》安东尼奥·维瓦尔第）

Six Flute Sonatas in A Minor, Sonata for Flute Suite No. 2 Johann Sebastian Bach
（《六首长笛奏鸣曲 a 小调无伴奏奏鸣曲 长笛第二号组曲》
约翰·塞巴斯蒂安·巴赫）

Five Flute Sonatas, Twelve Flute Fantasia Georg Philipp Telemann
（《五首长笛奏鸣曲》《12 首长笛幻想曲》格奥尔格·菲利普·泰勒曼）

Flute Concerto in D Major Behrs   （《D 大调长笛协奏曲》贝尔斯）

Flute Concerto in C Major, G Major, D Major   Wolfgang Amadeus Mozart
（《C 大调、G 大调、D 大调长笛协奏曲》沃尔夫冈·阿玛多伊斯·莫扎特）
Shepherd Boy Piccolo   He Luting   （《牧童短笛》贺绿汀）
Meditation   He Luting   （《幽思》贺绿汀）
Lullaby   He Luting   （《摇篮曲》贺绿汀）
The Setting Sun Rustling Drum   Tan Mizi   （《夕阳萧鼓》谭密子）
The Sun Shines Brightly on the Tianshan Mountains   Huang Huwei
                                                      （《阳光灿烂照天山》黄虎威）

Notes on the text

1. head joint 笛头
2. court music 宫廷音乐，是古典音乐的一种，流行于宫廷中。
3. secular music 世俗音乐（指非宗教的音乐）
4. traverso 单键长笛
5. Boehm flute 波姆式长笛
6. Theobald Boehm 特奥巴尔德·波姆 (1973—1881)，出生于德国慕尼黑的奥尔塞默芮克，德国长笛演奏家、乐器改革家、作曲家，被称为管乐史最有贡献的管乐家。
7. diaphragmatic breathing 腹式呼吸
8. circular breathing 循环呼吸

Terms

crown	/kraʊn/	笛头塞；软木塞子
embouchure hole	/ˌɑːmbʊˈʃʊr həʊl/	吹孔
lip plate (or embouchure plate)	/lɪp pleɪt/	嘴唇贴盘
trill key	/trɪl kiː/	颤音键
diaphragmatic breathing	/daɪəfræɡˈmætɪk ˈbriːðɪŋ/	腹式呼吸
circular breathing	/ˈsɜːrkjələr ˈbriːðɪŋ/	循环呼吸
overblow	/ˌoʊvəˈbloʊ/	超吹
hum at the same time	/hʌm ət ðə seɪm taɪm/	同时哼唱
single tonguing	/ˈsɪŋɡl ˈtʌŋɪŋ/	单吐
double tonguing	/ˈdʌbl ˈtʌŋɪŋ/	双吐
triple tonguing	/ˈtrɪpl ˈtʌŋɪŋ/	三吐
vibrato	/vɪˈbrɑːtəʊ/	（唇）颤音
flutter tonguing	/ˈflʌtər ˈtʌŋɪŋ/	颤舌音；花舌音
trill	/trɪl/	（指）颤音

glissando	/glɪˈsændəʊ/	滑音
legato	/lɪˈɡɑːtəʊ/	连奏

Exercises

I. Comprehension questions

1. Is the flute an end-blown wind instrument?
2. What was the original transverse flute called in Europe about 800 AD?
3. What was the Baroque flute called? Describe the Baroque flute.
4. What is the difference between the flute with C foot and the one with B foot?
5. What is the difference between the flute with inline G key and the one with offset G key?

II. Translating useful expressions

1. 长笛是无簧、横吹（侧吹）木管乐器。
2. 长笛在高音谱号记谱，是现代管弦乐和室内乐中主要的高音旋律乐器。
3. 长笛由笛头、笛身和笛尾三个部分组成。
4. The tone color of the flute is lively and bright in the high register, and beautiful and pleasant in the low register.
5. Diaphragmatic breathing and circular breathing are two main breathing techniques for the flute players to use.
6. Boehm's major innovation to flutes were the change to metal instead of wood, large straight tube bore, the size and placement of tone holes, "parabolic" tapered head joint bore, very large tone holes covered by keys, and the linked key system, which simplified fingering.

III. Brainstorm

In China there are many varieties of *dizi* (笛子 or Chinese flute), with different sizes, structures (with or without a resonance membrane) and number of holes (from 6 to 11) and intonations (different keys). Please make a research on the commonalities and differences between the Chinese flute and the Western concert flute, and present in class.

Clarinet

The clarinet (*or clarinetto*) is a family of woodwind instrument, and is a transposing instrument. The clarinet has a beak-shaped[1] mouthpiece with a single reed, a straight, cylindrical tuba with an almost cylindrical bore, and a flared bell. Clarinets have the largest pitch range of common woodwinds and they are known as the "orator" in the orchestra and the "dramatic soprano" in the woodwind. The most common clarinet is the Bb clarinet. The clarinet is widely used as a solo instrument, and has been a flexible instrument, used in concert bands, military bands, marching bands, klezmer, jazz, and other styles.

I. Construction

The vast majority of clarinets used by professionals are made from African blackwood (or grenadilla). The clarinet is composed of five parts: the mouthpiece with a single reed; the barrel, a piece of tube that bulges like a barrel; the upper joint (left-hand joint); the lower joint (right-hand joint) and the funnel-shaped[2] bell.

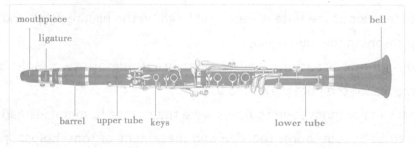

Mouthpiece and reed

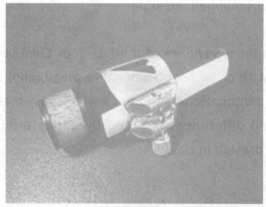

clarinet reed, mouthpiece, and vandoren ligature

The clarinet uses a single reed made from the cane of Arundo donax, a type of grass. The reed is attached to the mouthpiece by the ligature, and the top half-inch or so of this assembly is held in the player's mouth. Reed and mouthpiece work together to determine ease of playability, pitch stability, and tonal characteristics.

Barrel

barrel of a B♭ soprano clarinet

The short barrel may be extended to fine-tune the clarinet. Additional compensation for pitch variation and tuning can be made by pulling out the barrel and thus increasing the instrument's length.

Upper joint and lower joint

 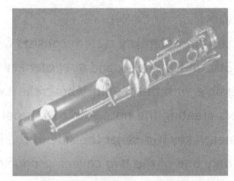

upper joint of a Boehm system clarinet   *lower joint of a Boehm system clarinet*

The main body of most clarinets is divided into the *upper joint*, the holes and most keys of which are operated by the left hand, and the *lower joint* with holes and most keys operated by the right hand. The body of a modern soprano clarinet is equipped with numerous *tone holes* of which seven (six front, one back) are covered with the fingertips, and the rest are opened or closed using a set of *keys*. These tone holes let the player produce every note of the chromatic scale. The cluster of keys at the bottom of the upper joint (protruding slightly beyond the cork of the joint) are known as the trill keys and are operated by the right hand. These give the player alternative fingerings that make it easy to play ornaments and trills. The entire weight of the smaller clarinets is supported by the right thumb behind the lower joint on what is called the thumb-rest.

Bell

bell of a B♭ soprano clarinet

The bell has a flared end. Contrary to popular belief, the bell does not amplify the sound; rather, it improves the uniformity of the instrument's tone for the lowest notes in each register. For the other notes, the sound is produced almost entirely at the tone holes, and the bell is irrelevant. On basset horns and larger clarinets, the bell curves up and forward and is usually made of metal.

Keywork

The current Böhm key system consists of generally 6 rings, on the thumb, 1st, 2nd, 4th, 5th, and 6th holes, and *a register key* (泛音键) just above the thumb hole, easily accessible with the thumb. Above the 1st hole, there is *a key* (A 键) that lifts two covers creating the note A in the throat register (high part of low register) of the clarinet. A key (G# 键)at the side of the instrument at the same height as the A key lifts only one of the two covers, producing G#, a semitone lower. The A key can be used in conjunction solely with the register key to produce A#/B♭.

II. Types of Clarinet Family

The clarinets of different sizes and pitches make up the largest woodwind family: from the (extremely rare) BBBb octo-contrabass[3] to the Ab piccolo clarinet[4]. Nowadays, the most common clarinet usually refers to the Bb soprano clarinet[5], which has a large range of nearly four octaves. The following are the most important sizes in the clarinet family, from the lowest to highest:

contrabass   bass   alto   Bb   C   Eb

1. Octocontrabass clarinet 最低低音单簧管 (BBBb 调)
2. Octocontralto clarinet 最低中音单簧管 (EEEb 调)
3. Contrabass clarinet 倍低音单簧管 (BBb/GG 调)
4. Contralto clarinet 倍中音单簧管 (EEb 调)
5. Bass clarinet 低音单簧管单簧管 (BB/AA 调)
6. Alto clarinet 中音单簧管 (Eb 调)
7. Basset Horn 巴塞特管 (F/G 调)
8. Soprano clarinet 高音单簧管 (Bb/A 调)
9. Sopranino clarinet 超高音单簧管 (Ab/Eb/D/C 调)

III. Characteristics in Range and Timbre

Because the clarinet has the largest pitch range of common woodwinds, the range of a clarinet can be divided into three distinct registers: the lowest register (or the *chalumeau* register); the middle register (or the *clarion* register), and the top register (or altissimo register), in which the middle register is the dominant range for most members of the clarinet family. All three registers have characteristically different sounds. The *chalumeau* register is rich and dark; the *clarion* register is brighter and sweet, while the *altissimo* register can be piercing and sometimes shrill.

The cylindrical bore is primarily responsible for the clarinet's distinctive timbre, which varies among its three main registers. The tone quality can vary greatly with the clarinetist, music, instrument, mouthpiece, and reed. The A and B♭ clarinets have nearly identical tonal quality, although the A typically has a slightly warmer sound. The tone of the E♭ clarinet is brighter and the bass clarinet has a characteristically deep, mellow sound, while the alto clarinet is similar in tone to the bass.

IV. Development

The clarinet has its roots in the early single-reed instruments or hornpipes used in Ancient Greece, Ancient Egypt, Middle East, and Europe since the Middle Ages, such as the albogue and alboka.

The modern clarinet developed from a Baroque instrument called the *chalumeau*[6] which had a single-reed mouthpiece and a cylindrical bore, and had eight finger holes, like a recorder, and two keys for its highest notes. Due to lacking a register key, the *chalumeau* could be mainly played in its fundamental register, with a limited range of about one and a half octave.

Based on the *chalumeau*, Johann Christoph Denner invented the first clarinet in Germany around 1700 by adding two finger keys and a better mouthpiece to a chalumeau. Denner's clarinet played well in the middle register with loud, shrill sound, so it was given the name *clarinetto*. These early clarinets did not play well in the lower register, so players continued to play the *chalumeau* for low notes. As clarinets improved, the *chalumeau* stopped using, but these notes remained as the chalumeau register.

Because original Denner clarinets had two keys, and could play a chromatic scale, various clarinet makers continued to add more keys to get improved tuning, easier fingerings, and a slightly larger range. The classical clarinet of Mozart's day typically had eight finger holes and five keys. Clarinets were soon accepted into orchestras, and by the time of Beethoven, the instrument was a standard fixture in the orchestra.

The next major development in the history of clarinet was the invention of the modern pad. In 1812, Iwan Muller developed a new type of pad that was covered in leather or fish bladder. The pad was airtight and let makers increase the number of pad-covered holes. Muller then designed a new type of clarinet with seven finger hole and 13 keys, which allowed the instrument to play in any key with near-equal ease.

The final development in the clarinet was made by a French player Hyacinthe Klose who adopted the Boehm flute key system to improve the clarinet, and devised a different arrangement of keys and finger holes for the instrument in 1839. The Boehm system added a series of rings and axles that made fingering easier, which greatly widen the tonal range of the clarinet. Today the Boehm system is used everywhere in the world except Germany and Austria. Germany and Austria still use a direct descendant of the Mueller clarinet known as Oehler system. The Mueller system was improved by Oscar Oehler who made the Mueller clarinet more delicate at about 1900.

Today, two systems are in use: the *German* (or the *Oehler*) *system*[7] and the *French* (or the *Boehm*) *system*[8]. The *Boehm clarinet* has seventeen keys (plus seven open tone holes, making twenty-four tone holes all together) and six ring keys, less than the Öhler clarinet, which contains at least twenty-two keys and five ring keys.

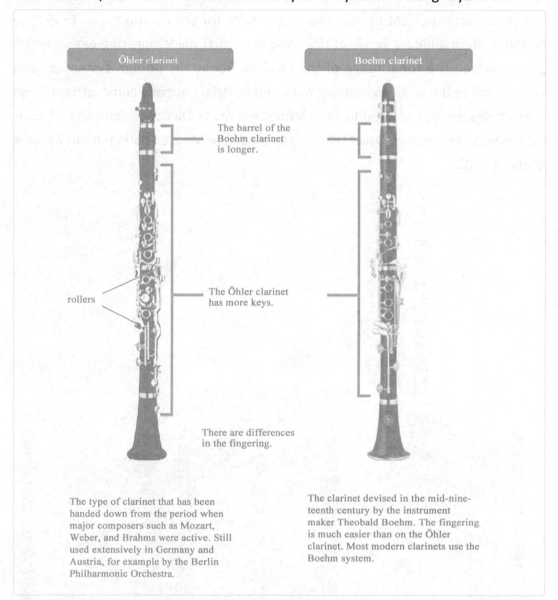

V. Performing Techniques

Embouchure

The formation of the mouth around the mouthpiece and reed is called the embouchure. When playing the clarinet, the reed is placed on the lower lip, which is pressed against the lower teeth while the upper teeth grip the mouthpiece on the closed side. When air is blown through the opening between the reed and the mouthpiece facing, the reed vibrates and produces the clarinet's sound.

Main performing techniques

Vibrato: Pitch and/or volume are produced by movements of the diaphragm, larynx and lips.

Flutter tonguing: The clarinetist articulates a lingual R (produced with the tip of the tongue) or a rolled (guttural) R against the palate.

Trill: As a typical playing technique, although the fingering of some trills makes them difficult to play, certain trills are made easier by trill keys. The higher the trill is, the more penetrating the sound.

Glissando: It is produced by partly covering the tone holes or by changes of fingering, playable over the whole compass, but with some limitations: glissandos which cross from one register to another are generally very difficult to play.

Arpeggio: The performance of broken chords is particularly characteristic of clarinet playing.

Legato, single tonguing, diaphragmatic breathing, circular breathing and large leaps are typical playing techniques on the clarinet.

VI. Musical Classics

Clarinet Quintet in A Major K. 581   Wolfgang Amadeus Mozart
　　　（《A 大调单簧管五重奏》K. 581 沃尔夫冈·阿玛多伊斯·莫扎特）
Clarinet Concerto in A Major K. 622   Wolfgang Amadeus Mozart
　　　（《A 大调单簧管协奏曲》K. 622 沃尔夫冈·阿玛多伊斯·莫扎特）
Clarinet Trio in B-Flat Major K498   Wolfgang Amadeus Mozart
　　　（《降 B 大调单簧管三重奏》K. 498 沃尔夫冈·阿玛多伊斯·莫扎特）
Clarinet Trio in A Minor, Op. 114   Johannes Brahms
　　　（《单簧管三重奏》Op. 114 约翰内斯·勃拉姆斯）
Clarinet Quintet in B Minor, Op. 115   Johannes Brahms
　　　（《B 小调单簧管五重奏》Op. 115 约翰内斯·勃拉姆斯）
Sonata for Clarinet and Piano No. 1 & 2, Op.120   Johannes Brahms
　　　（《单簧管奏鸣曲》Op.120-1,2 约翰内斯·勃拉姆斯）
Concerto for Clarinet and Orchestra No.1 in F Minor, Op. 73   Carl Maria von Weber
　　　（《单簧管第一协奏曲》Op.73 卡尔·马利亚·冯·韦伯）

Clarinet Concerto No. 2 in E Flat Major, Op. 74　Carl Maria von Weber
（《单簧管第二协奏曲》Op.74 卡尔·马利亚·冯·韦伯）

Fantasy Pieces for Clarinet and Piano in A Minor, Op. 73　Robert Schumann
（《单簧管与钢琴幻想曲》Op. 73 罗伯特·舒曼）

Three Romances Op. 94　Robert Schumann
（《三首浪漫曲》Op. 94 罗伯特·舒曼）

Marchenzahlungen for Clarinet, Viola and Piano, Op.132　Robert Schumann
（《单簧管、中提琴与钢琴奏鸣曲 - 童话》Op. 132 罗伯特·舒曼）

Introduction Theme and Variations　Gioachino Rossini
（《引子主题与变奏》焦阿基诺·安东尼奥·罗西尼）

Fantasy on "La Traviata"　Giuseppe Verdi　（《茶花女幻想曲》居塞比·威尔第）

N3 Clarinet Concerto　Louis Spohr　（《第三单簧管协奏曲》路易斯·施波尔）

N4 Clarinet Concerto　Louis Spohr　（《第四单簧管协奏曲》路易斯·施波尔）

2 Koncertstuck for Basset Horn, Clarinet and Piano No.1 in F Minor, Op.113
Jakob Ludwig Felix Mendelssohn Bartholdy
（《为单簧管、巴赛特管与钢琴所作的音乐会小品》Op.113 雅科布·路德维希·费利克斯·门德尔松·巴托尔迪）

2 Koncertstuck for Basset Horn, Clarinet and Piano No.2 in D Minor Op.114
Jakob Ludwig Felix Mendelssohn Bartholdy
（《为单簧管、巴赛特管与钢琴所作的音乐会小品》Op.114 雅科布·路德维希·费利克斯·门德尔松·巴托尔迪）

Première Rhapsodie for Clarinet, Piano, and Orchestra　Achille-Claude Debussy
（《第一狂想曲》阿西尔 - 克劳德·德彪西）

Notes on the text

1. the beak-shaped 鸟嘴形
2. funnel-shaped 漏斗形的
3. BBBb octo-contrabass 最低低音单簧管 (BBBb 调)
4. Ab piccolo clarinet Ab 高音单簧管
5. Bb soprano clarinet 降 B 调高音单簧管
6. chalumeau 沙吕莫管, 芦笛
7. Oehler system 德式或欧勒体系
8. Boehm system 法式或波姆体系

Lesson Eight
Woodwind Instruments

🎻 Terms

clarinet	/ˌklærəˈnet/	单簧管（又称竖笛、黑管）
cylindrical	/səˈlɪndrɪkl/	圆柱形的
clarinetist	/ˌklærəˈnetɪst/	单簧管演奏者
octave	/ˈɑːktɪv/	八度音阶
mouthpiece	/ˈmaʊθpiːs/	吹嘴；吹口；笛头
ligature	/ˈlɪɡətʃər/	束圈
barrel	/ˈbærəl/	调节管；二节管；脖管
register key	/ˈredʒɪstər kiː/	泛音键
embouchure	/ˌɑːmbʊˈʃʊr/	吹口；口型
compass	/ˈkʌmpəs/	音域
fingering	/ˈfɪŋɡərɪŋ/	指法
overblow	/ˌoʊvəˈbloʊ/	超吹
overblown note	/ˌoʊvərˈbloʊn noʊt/	吹出泛音
fundamental	/ˌfʌndəˈmentl/	基音
single tonguing	/ˈsɪŋɡl ˈtʌŋɪŋ/	单吐
flutter tonguing	/ˈflʌtər ˈtʌŋɪŋ/	花舌音；颤舌音
vibrato	/vɪˈbrɑːtoʊ/	（唇）颤音
trill	/trɪl/	（指）颤音
glissando	/ɡlɪˈsændoʊ/	滑音
arpeggio	/ɑːrˈpedʒioʊ/	琶音
legato	/lɪˈɡɑːtoʊ/	连奏
run	/rʌn/	急奏
diaphragmatic breathing	/daɪəfræɡˈmætɪk ˈbriːðɪŋ/	腹式呼吸

Exercises

I. Comprehension questions

1. How many reeds does a clarinet have?
2. How many parts is a clarinet composed of?
3. Who adopted the Boehm flute key system to improve the clarinet in 1839?
4. How did Hyacinthe Klose improve the former clarinet?
5. How does a clarinet produce the sound?

II. Translating useful expressions

1. 单簧管有一个鸟嘴型的吹口、直的圆柱形的管体和一个漏斗形的喇叭口。
2. 目前，我国主要使用23键的波姆式单簧管。
3. 单簧管在木管乐器中音域最宽，并有极具特色的三个声区。
4. The clarinet is a family of woodwind instrument, and is a transposing instrument.
5. In 1812, Iwan Muller developed a new type of pad that was covered in leather or fish bladder. The pad was airtight and let makers increase the number of pad-covered holes.
6. The main body of most clarinets is divided into the upper joint, the holes and most keys of which are operated by the left hand, and the lower joint with holes and most keys operated by the right hand.

III. Brainstorm

The most common system of keys of the clarinet was named the Böhm system by its designer Hyacinthe Klosé in honour of flute designer Theobald Böhm, but it is not the same as the Böhm system used on flutes. Newton once said "*If I see farther than others, it is because I stand on the shoulders of giants*". Many giants, like Newton and Hyacinthe Klosé, are quite humble (谦逊的) about their success. Being a humble person, what can we learn from Newton and Hyacinthe Klosé? Please give a specific example to discuss what qualities make a person successful.

Lesson Eight
Woodwind Instruments

Saxophone

The saxophone (referred to colloquially as the sax) is a type of single-reed woodwind instrument with a conical body, usually made of brass. Its sound is produced when air is blown through the instrument causing the reed to vibrate, and is amplified as it travels through the instrument's main body. Saxophones are made in various sizes and are almost always treated as transposing instruments.

The saxophone is used in a wide range of musical styles including classical music (such as concert bands, chamber music, solo repertoire, and occasionally orchestras), military bands, marching bands, jazz (such as big bands and jazz combos), and contemporary music. The saxophone is also used as a solo and melody instrument or as a member of a horn section in some styles of rock and roll and popular music.

I. Construction

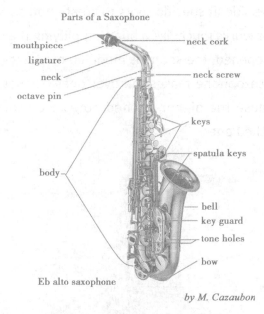

by M. Cazaubon

Saxophones consist of numerous parts and pieces which are made separately and then assembled. The main parts of the saxophone have a mouthpiece, conical metal tube, and keys, and the main body of the saxophone is primarily made from brass.

Mouthpiece and reed

The saxophone uses a single-reed mouthpiece similar to that of the clarinet. Each size of saxophone (alto, tenor, etc.) uses a different size of reed and mouthpiece. Mouthpiece design has a profound impact on tone. The two main parts of the mouthpiece affect tone: the tone chamber and the lay (or the facing) which is the opening between the mouthpiece's reed and its tip. Mouthpieces are typically marked with a letter or number to denote the width of the lay.

Most saxophonists use reeds made from Arundo donax[1] cane, but since the middle of the twentieth century some have been made of fiberglass or other composite materials. The reed can be made soft or hard depending on the desire of the musician. The ligature is the part that holds the reed on the mouthpiece. It attaches to the mouthpiece with screws.

Tube

The crook is the part that joins the mouthpiece and the main instrument body. At the top of it is a cork which is important for tuning the instrument. The tone changes depending on where the mouthpiece is positioned on the cork. The other end of the crook is a metal joint that fits into the main body of the saxophone. It connects with a screw to keep the crook in place.

The saxophone tube is a long, metal tube which steadily gets wider at one end. It has holes drilled in the side at specific spots to create notes. When all the holes are closed, the instrument works much like a bugle amplifying the sound of the vibrating reed. When a hole is opened, the sound is modified producing a different note. The conical shape of the saxophone makes the overtones octaves, and this also makes fingering easier because the higher pitched notes are produced with the same fingering as lower pitched ones.

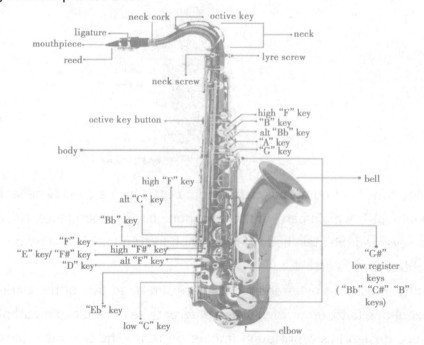

Key

Saxophone keys have two types: closed standing and open standing. Closed standing keys are those that are held closed by a spring when the instrument is not being played. When the key is pressed, the hole it covers is opened. Open standing keys are held open by a spring and close when the key is pressed. The octave key raises the pitch of the lower notes by one octave. The lowest possible note, with all of the pads closed, is the B♭ below middle C. Each key has a pad on its end which provides an airtight seal on the hole.

The pitch of a saxophone is controlled by opening or closing the tone holes along the body of the instrument to change the length of the vibrating air column. Most of the keys are operated by the player's fingers, but some are operated using the palm or the side of a finger.

II. Types

The typical saxophone is a single reed instrument made from brass with a curved bottom. The modern saxophone family consists entirely of B♭ and E♭ instruments. From highest pitch to lowest pitch, there are the E♭ sopranino, B♭ soprano, E♭ alto, B♭ tenor, E♭ baritone, and B♭ bass saxophones. The saxophones in widest use are the B♭ soprano, E♭ alto, B♭ tenor, and E♭ baritone. The E♭ sopranino and B♭ bass saxophone are typically used in larger saxophone choir settings, when available.

Soprano and sopranino saxophones are usually constructed with a straight tube with a flared bell at the end, although some are made in the curved shape of the other saxophones. Alto and larger saxophones have a detachable curved neck and a U-shaped bend (the bow) that directs the tubing upward as it approaches the bell. The baritone, bass, and contrabass saxophones accommodate the length of the bore with extra bends in the tube.

The fingering system for the saxophone is similar to the systems used for the oboe, the Boehm-system[2] clarinet, and the flute. Saxophone music is written in treble clef and all saxophones use the same key arrangement and fingerings, enabling players to switch between different types of saxophones fairly easily.

III. Development

Early development and adoption

The saxophone was designed around 1840 by Adolphe Sax[3], a Belgian instrument maker, flautist, and clarinetist. Born in Dinant and originally based in Brussels, he moved to Paris in 1842 to establish his musical instrument business. Before working on the saxophone, he made several improvements to the bass clarinet by improving its keywork and acoustics and extending its lower range. Sax was also a maker of the ophicleide, a large conical brass instrument in the bass register with keys similar to a woodwind instrument. His experience with these two instruments allowed him to develop the skills and technologies needed to make the first saxophones.

As an outgrowth of his work improving the bass clarinet, Sax began developing an instrument with the projection of a brass instrument and the agility of a woodwind. The first saxophone Sax created was a large, bass saxophone with a single-reed mouthpiece and conical brass body. Having constructed saxophones in several sizes in the early 1840s, Sax applied for, and received, a 15-year patent for the instrument on 28 June 1846.

In the 1840s and 1850s, Sax's invention gained use in small classical ensembles (both all-saxophone and mixed), as a solo instrument, and in French and British military bands. Saxophone method books were published and saxophone instruction was offered at conservatories in France, Switzerland, Belgium, Spain, and Italy. The saxophone was used experimentally in orchestral scores, but never came into widespread use as an orchestral instrument.

After an early period of interest and support from classical music communities in Europe, their interest in the instrument waned in the late nineteenth century. Saxophone teaching at the Paris Conservatory was suspended from 1870 to 1900 and classical saxophone repertoire stagnated during that period, but it was during this same period that the saxophone began to be promoted in the United States, many American musicians and saxophonists made great efforts to broaden the adoption of the saxophone.

Early 20th-century growth and development

While the saxophone remained marginal and regarded mainly as a novelty instrument in the classical music world, many new musical niches were established for it during the early decades of the twentieth century. Its early use in vaudeville and ragtime bands[4] around the turn of the century laid the groundwork for its use in dance orchestras and eventually jazz.

The use of the saxophone for more dynamic and more technically demanding playing styles added incentive for improvements in keywork and acoustic design. Early saxophones had two separate octave keys operated by the left thumb to control the two octave vents required on alto or larger saxophones. A substantial advance in keywork around the turn of the century was development of mechanisms, by which the left thumb operates the two octave vents with a single octave key. Ergonomic design of keywork evolved rapidly during the 1920s and 1930s. The front F mechanism supporting alternate fingerings for high E and F, and stack-linked G♯ key action became standard during the 1920s, followed by improvements to the left hand table key mechanisms controlling G♯ and bell keys. New bore designs during the 1920s and 1930s resulted from the quest for improved intonation, dynamic response and tonal qualities.

Modern saxophone emerges

The modern layout of the saxophone emerged during the 1930s and 1940s, first with right-side bell keys introduced on baritones, then on altos and tenors. The mechanics of the left hand table were revolutionized by Selmer[5] in 1936, and thirty to forty years later this final Selmer layout was nearly universal on all saxophone models. From the 1940s on, many conservatoires in the world established study of the saxophone as a classical instrument.

IV. Preforming Techniques

Saxophone technique refers to the physical means of playing the saxophone. It includes how to hold the instrument, how the embouchure is formed and the airstream produced, tone production, hands and fingering positions, and a number of other aspects.

Embouchure

Saxophone embouchure is the position of the facial muscles and shaping of the lips to the mouthpiece when playing a saxophone. Playing technique for the saxophone can derive from an intended style (classical, jazz, rock, funk, etc.) and the player's idealized sound. The design of the saxophone allows for a wide variety of different approaches to sound production. However, there is a basic underlying structure to most techniques.

The most common saxophone embouchures in modern music use are variants of the single-lip embouchure, in which the mouthpiece position is stabilized with firm pressure from the upper teeth resting on the mouthpiece (sometimes padded with a thin strip of rubber known as a "bite-pad" or "mouthpiece-patch"). The lower lip is supported by the buccinator and chin muscles and rests in contact with the lower teeth, making contact with the reed. The mouthpiece is generally inserted with the beak not taken more than halfway into the player's mouth.

Vibrato is made by using the jaw, tongue, lip, throat and diaphragm, so generally, there are jaw vibrato, lip vibrato, throat vibrato and diaphragm vibrato.

The jaw motions required for vibrato can be simulated by saying the syllables "wah-wah-wah" or "tai-yai-yai."

The lip vibrato, which is often confused with the jaw vibrato, is produced by moving the lips in something like a "wa-wa-wa---" motion. However, this is more difficult to control, as it causes a greater disturbance to the basic embouchure.

The throat vibrato, which is seldom used any more, is a type of "spasm" generated by tensing the throat muscles, and results in a sort of "quiver."

The diaphragm vibrato, sometimes called "breath vibrato", is predominantly an intensity vibrato. It is induced by a changing of the rate of the air pressure on the reed, and accomplished by moving the abdominal muscles, which in turn put pressure on the diaphragm, much as one would say "huh-huh-huh--."

Growling is a technique used whereby the saxophonist sings, hums, or growls, using the back of the throat while playing. This causes a modulation of the sound, and results in a gruffness or coarseness of the sound. It is rarely found in classical or band music, but is often utilized in jazz, blues, rock 'n' roll, and other popular genres.

Glissando is a pitch technique where the saxophonist bends the pitch of the note using voicing (tongue and embouchure placement) to move to another fingered note.

Multiphonics is the technique of playing more than one note at once. A special fingering combination causes the instrument to vibrate at two different pitches alternately, creating a warbling sound. A similar effect can also be created by 'humming' while playing a note.

The use of overtones involves fingering one note but altering the air stream to produce another note which is an overtone of the fingered note. For example, if low B♭ is fingered, a B♭ one octave above may be sounded by manipulating the air stream.

Voicing is the technique of manipulating the air stream to obtain various effects. Voicing technique involves varying the position of the tongue and throat, causing the same amount of air to pass through either a more or less confined oral cavity.

This causes the air stream to either speed up or slow down, respectively.

Slap tonguing creates a "popping" or percussive sound. A slap may be notated either pitched, or non-pitched.

Flutter-tonguing can give a rolling R sound with the tone played.

Overblow is a technique used while playing a wind instrument which, primarily through manipulation of the supplied air (versus, e.g., a fingering change or operation of a slide), causes the sounded pitch to jump to a higher one.

Circular Breath is a technique that can let players produce a continuous tone without interruption. This is accomplished by breathing in through the nose while simultaneously pushing air out through the mouth using air stored in the cheeks.

Double and Triple Tonguing is a technique that involves the tip and back of the tongue. The technique involves emulating the sounds "ta-ca" or "ti-gui", both of which employ the tip and back of the tongue. This allows the player to tongue-articulate at twice the speed that the single-tonguing technique allows.

V. Musical Classics

Boléro   Maurice Ravel   （《波莱罗舞曲》 莫里斯·拉威尔）

Pictures at an Exhibition (Ravel version)   Mojeste Petrovich Mussorgsky
　　（《展览会之画》/《图画展览会》 莫杰斯特·彼得罗维奇·穆索尔斯基）

L'Arlésienne   Georges Bizet   （《阿莱城的姑娘》乔治·比才）

Romeo and Juliet   Sergei Prokofiev
　　（《罗密欧与朱丽叶》 谢尔盖·普罗科菲耶夫）

Rhapsody in Blue   George Gershwin   （《蓝色狂想曲》乔治·格什温）

West Side Story   Leonard Bernstein   （《西区故事》雷昂纳德·伯恩斯坦）

Going Home   Kenny G   （《回家》肯尼·基）

Forever In Love   Kenny G   （《永沐爱河》肯尼·基）

Jasmie Flower   Kenny G   （《茉莉花》肯尼·基）

Notes on the text

1. Arundo donax 芦竹
2. Boehm-system 波姆体系
3. Adolphe Sax 阿道夫·萨克斯 (1814—1894)，比利时人、萨克斯风发明者。
4. ragtime bands 爵士乐队
5. Selmer 塞尔曼，是萨克斯品牌。

Terms

reed	/ri:d/	簧片
mouthpiece	/ˈmaʊθpi:s/	吹嘴
ligature	/ˈlɪgətʃər/	卡子；吹口束圈
neck/ crook	/nek/, /krʊk/	弯脖；吹管；颈管；共鸣管；琴颈
octave pin	/ˈɑːktɪv pɪn/	泛音键联动杆
bow	/baʊ/	弯管（U 形管）
key guard	/kiː gɑːrd/	音键保护框架；保护网
bell	/bel/	喇叭口
spatula key	/ˈspætʃələ kiː/	桌键
neck screw	/nek skruː/	吹管固定螺丝
octave key	/ˈɑːktɪv kiː/	高八度按键
cork	/kɔːrk/	软木塞
vibrato	/vɪˈbrɑːtəʊ/	颤音
glissando	/glɪˈsændəʊ/	滑音
growling	/ˈgraʊlɪŋ/	咆哮音；爆破音
voicing	/ˈvɔɪsɪŋ/	音色
slap tonguing	/slæp ˈtʌŋɪŋ/	弹舌
multiphonics	/mʌltiˈfɑːnɪks/	复合音
overtone	/ˈoʊvərtoʊn/	泛音
flutter-tonguing	/ˈflʌtər ˈtʌŋɪŋ/	花舌音；舌颤音
overblow	/ˌoʊvəˈbloʊ/	超吹
circular breathing	/ˈsɜːrkjələr ˈbriːðɪŋ/	循环呼吸
double tonguing	/ˈdʌbl ˈtʌŋɪŋ/	双吐音
triple tonguing	/ˈtrɪpl ˈtʌŋɪŋ/	三吐音

Exercises

I. Comprehension questions

1. Who invented the saxophone around 1840?
2. What are the shapes of the saxophones?
3. What types of saxophones are in widest use?
4. When was saxophone teaching at the Paris Conservatory suspended and classical saxophone repertoire stagnated?

5. What were the growth and development of the saxophone in the early 20th century?

II. Translating useful expressions

1. 萨克斯是一种单簧木管乐器，并有圆锥型的管身。
2. 萨克斯有两种键：闭孔键和开孔键。
3. 颤音、滑音和超吹都是萨克斯很有特色的一种技巧。
4. Playing technique for the saxophone can derive from an intended style (classical, jazz, rock, etc.) and the player's idealized sound.
5. The modern layout of the saxophone emerged during the 1930s and 1940s, first with right-side bell keys introduced on baritones, then on altos and tenors.
6. The pitch of a saxophone is controlled by opening or closing the tone holes along the body of the instrument to change the length of the vibrating air column.

III. Brainstorm

Adolphe Sax is not only an instrument maker, flautist, and clarinetist, but also the inventor of the saxophone. Bartolomeo Cristofori, who is an expert harpsichord maker and is well-acquainted with the knowledge on stringed keyboard instruments, invents the first piano. Theobald Boehm is a great flutist, composer, acoustician, and silversmith, but his greatest contribution is the innovation to flutes. Please research the commonalities of the three giants, and then discuss what you have gained.

Oboe

The oboe is a double reed woodwind instrument, and is not a transposing instrument. The most common oboe plays in the treble range, and its commonly accepted range has over two and a half octaves. A soprano oboe measures roughly 65 cm (25.5 in) long, with metal keys, a conical bore and a flared bell. When the word oboe is used alone, it is generally taken to mean the treble instrument rather than other instruments of the family, such as the bass oboe, the cor anglais[1] (English horn), or oboe d'amore[2].

The oboe is the main melody instrument in the orchestras. Because of its secure and penetrating sound, the oboe is not only the ideal instrument to tune the orchestra with its distinctive "A", but also is easier to blends well with other instruments in large ensembles. Due to its soft and beautiful timbre, the oboe is suitable for expressing pastoral theme and melancholy and lyrical mood. It is known as "lyrical soprano" in the instruments. Today, the oboe is commonly used as orchestra or solo instrument in symphony orchestras, concert bands and chamber ensembles. The oboe is especially used in chamber music, film music, some genres of folk music, and is occasionally heard in jazz, rock, pop, and popular music.

I. Construction

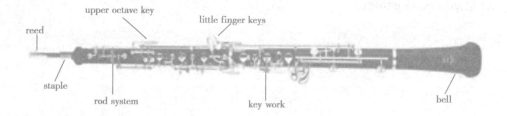

Oboes are usually made of wood, but may also be made of synthetic materials, such as plastic, resin or hybrid composites. The oboe consists of three sections: top (upper joint), middle (lower joint), and bell. The keys vary in function and the keys on the back are operated by the thumb to help to play the high octaves.

Lesson Eight
Woodwind Instruments

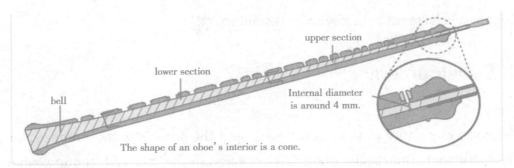

a cross-section of an oboe

The inside of an oboe is very narrow, indeed, with the internal diameter of the upper portion of the upper section measuring only about 4 mm. The diameter grows wider the closer to the bell, forming a cone. The narrow blowing hole allows a little bit of the player's breath to enter, so the oboe is very difficult to play.

Reed

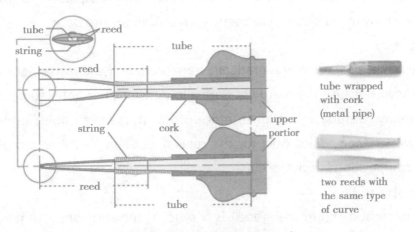

the structure of the double reeds

The two reeds are placed face-to-face and are strapped to the metal pipe with strings. A piece of cork wrapped around part of the reed is inserted into the upper section of the instrument. The two reeds are curved subtly, and so there is a slight gap in the center when the two ends are stuck together. This slight gap allows the player's puff to pass through. During playing, the reeds undergo minute vibrations, the gap between the reeds repeatedly closing and opening.

The reed is vital to the oboe. Like clarinet, saxophone, and bassoon reeds, oboe reeds are made from Arundo donax. The reed is considered the part of oboe that makes the instrument so difficult because the individual nature of each reed means that it is hard to achieve a consistent sound. Slight variations in temperature, altitude, weather, and climate can affect the sound of the reed, so oboists often prepare several reeds to achieve a consistent sound.

II. Development

Early

The oboe originated in the middle ages, and the shawm is the predecessor of the instrument. The regular oboe first appeared in the mid-17th century, when it was called a hautbois (in which *haut* means "high", "loud"; *bois* means "wood", "woodwind"). The hautbois quickly spread throughout Europe, and it was the main melody instrument in early military bands, until it was succeeded by the clarinet.

The baroque oboe is generally made of boxwood and has only two keys at the bottom: a C key and a E flat key, and the rest are just holes. By the early 18th century, the oboe had become a standard fixture in baroque orchestras and was used extensively by baroque composers, such as Handel and Bach.

Classical era

The bore of classical oboe was gradually narrowed, and the instrument was added to seven keys. Because the narrower bore allows the higher notes to be more easily played, many composers began to more often utilize the oboe's upper register in their works. Many solos exist for the regular oboe in chamber, symphonic, and operatic compositions from the Classical era.

Wiener oboe

The Wiener oboe (Viennese oboe) is a type of modern oboe that retains the essential bore and tonal characteristics of the historical oboe. The Wiener oboe with thirteen keys was first developed by Josef Hajek in the late 19th century. It has a wider internal bore, a shorter and broader reed and the fingering-system is very different than the conservatoire oboe. Its great advantages are the ease of speaking, even in the lowest register. The Wiener oboe can be played very expressively and blends well with other instruments.

Conservatoire oboe

By using the theory of the Boehm flute, the Triebert family of Paris designed a series of increasingly complex yet functional key systems in 19th century. In 1882, the Boehm system oboe was successfully used by Georges Cillet in Paris conservatory, so it has been called conservatoire ("conservatory" in the US) oboe since then. The Boehm system oboe was never in common use until F. Loree of Paris made further improvement to it in the 20th century. Minor improvements to the bore and key work have continued through the 20th century, but there has been no fundamental change to the general characteristics of the oboe for several decades.

Modern oboe

The modern oboe has an extremely narrow conical bore and has about 48 pieces of keywork. There are mainly two types of modern oboes: semi-automatic oboes and full-automatic oboes, in which the first type is more common use.

The standard oboe has several siblings of various sizes and playing ranges. They are the cor Anglais (English horn), *piccolo oboe, oboe d'amore, bass oboe* (baritone oboe) and *hecklphone*[3]. The English horn, as the tenor (alto) member of the family, is the widely known and used today.

III. Performing Techniques

The sound of the oboe is produced by blowing into the reeds at a sufficient air pressure, causing them to vibrate the air column, so the player's playing techniques will be influenced by fingers, breath, embouchure, tongue and throat. The main playing techniques are as the following:

Breathing: Like other wind instruments, the oboe must use diaphragmatic breathing and control the air pressure through the abdomen. The abdomen is also an important part to control the vibrato effect.

Slur or legato: No tone stops between the notes, and the sound remains clear and continuous throughout the frequency change.

Glissando: It is quite hard to do on the oboe. There are two types of glissandos: short glissando and long glissando. The short glissando is done by using the mouth muscles, while the long one by releasing your fingers gradually off the keys.

Vibrato, floral (flutter) tonguing, single tonguing, double tonguing, triple tonguing, staccato and trill are still commonly used.

IV. Musical Classics

Oboe Concerto in C Major   Wolfgang Amadeus Mozart
　　　　　　　（《C大调双簧管协奏曲》沃尔夫冈·阿玛多伊斯·莫扎特）
Peter and the Wolf   Sergei Prokfiev　　（《彼得与狼》谢尔盖·普罗科菲耶夫）

The Swan Lake　Peter Ilyich Tchaikovsky

（《天鹅湖》彼得·伊利奇·柴可夫斯基）

Drei Romanzen for Oboe and Piano(op.94)　Robert Schuman

（《双簧管和钢琴浪漫曲》罗伯特·舒曼）

Returing from Grazing at Dusk　Xin Huguang　　　　（《黄昏牧归》辛沪光）

Notes on the text

1. cor Anglais (English horn) 英国管
2. oboe d'amore 抒情双簧管
3. hecklphone 黑克尔管

Terms

oboist	/ˈəʊbəʊɪst/	双簧管吹奏者
the upper joint	/ðə ˈʌpər dʒɔɪnt/	上节管
the lower joint	/ðɪ ˈləʊər dʒɔɪnt/	下节管
keywork	/kiː wɜːrk/	按键
rod system	/rɑːd ˈsɪstəm/	连杆系统
staple	/ˈsteɪpl/	哨座，嘴套
octave key	/ˈɑːktɪv kiː/	八度音键；高音键
trill key	/trɪl kiː/	颤音键
little finger key	/ˈlɪtl ˈfɪŋgər kiː/	小指键
reed	/riːd/	簧片
cork	/kɔːrk/	软木塞
slur	/slɜːr/	连奏；圆滑音
legato	/lɪˈgɑːtəʊ/	连音
glissando	/glɪˈsændəʊ/	滑音
vibrato	/vɪˈbrɑːtəʊ/	（唇）颤音
floral tonguing	/ˈflɔːrəl ˈtʌŋɪŋ/	花舌音
staccato	/stəˈkɑːtəʊ/	断音
trill	/trɪl/	（指）颤音

Lesson Eight
Woodwind Instruments

Exercises

I. Comprehension questions

1. When was the Wiener oboe with thirteen keys first developed by Josef Hajek?
2. How many sections does the oboe consist of? Name them.
3. What theory was the conservatoire oboe based on to develop the oboe?
4. What slight variations can have an effect on the sound of the reed?
5. How are the oboe reeds designed to produce the sound?

II. Translating useful expressions

1. 普通双簧管在高音区演奏，是交响乐队的主要旋律乐器。
2. 由于双簧管有很稳且极具穿透力的声音，所以它是管弦乐队里理想的调音基准乐器。
3. 与其他管乐一般，吹奏双簧管必须使用腹式呼吸。
4. By the early 18th century, the oboe had become a standard fixture in baroque orchestras and was used extensively by baroque composers.
5. The inside of an oboe is very narrow, indeed, with the internal diameter of the upper portion of the upper section measuring only about 4 mm.
6. The reed is vital to the oboe. The reed is considered the part of oboe that makes the instrument so difficult because the individual nature of each reed means that it is hard to achieve a consistent sound.

III. Brainstorm

The oboe has commonly used as an orchestra or solo instrument in symphony orchestras, concert bands and chamber ensembles since the Classical Era, and many great composers utilize the oboe in their works. Take the oboe solo from Tchaikovsky's Ballet *Swan Lake* as an example, to discuss the timbre characteristic of the oboe and what you can gain from the oboe solo passage in *Swan Lake*.

Bassoon

The bassoon is a bass woodwind instrument with double reed, and is the largest member of the woodwind family. As a non-transposing instrument, the bassoon has three and a half octaves, and typically its music is written in the bass and tenor clefs, and occasionally the treble. The bassoon is composed of six pieces, and is usually made of wood or synthetic plastic. The modern bassoon exists in two distinct primary forms: Buffet[1] (or French) and Heckel[2] (or German) systems. Because of its distinctive tone color, wide range, variety of character, and agility, the bassoon is widely used in orchestral, concert band, and chamber, and is occasionally heard in pop, rock, and jazz.

I. Construction

The bassoon has a long conical tube[3], and the overall height of is 1.34m tall, but the total sounding length is 2.54 m because tube is doubled back on itself. The bassoon is composed of six main pieces, including the reed, the bell, the bass joint (or long joint), the boot joint(or butt/double joint), the wing joint (or tenor joint), and the bocal (or crook).

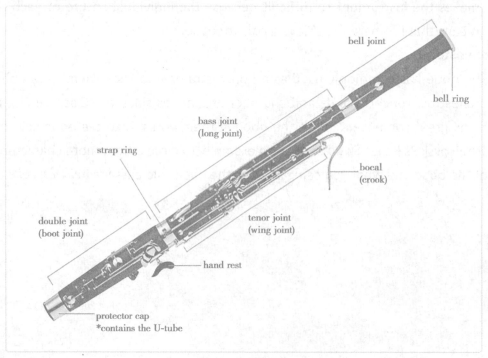

Lesson Eight
Woodwind Instruments

Reed (簧片 / 哨片): Modern bassoon reeds, made of Arundo donax, are often made by professional players themselves. The reed is consisted of two pieces of bamboo cane (芦竹) wrapped together with wire and thread, and bassoon reeds are usually around 5.5 cm (2.2 inch) in length. When you blow into the reed, the two pieces of cane vibrate against each other, and then the sound is produced.

reed

Bocal (or crook): A narrow, curved metal tube connects the double-reed mouthpiece to the wood body. The bassoon's pitch can not only be altered by the use of bocals of differing length, but also can be grossly adjusted by pushing the bocal in or out slightly.

Wing joint (or tenor joint): The wing joint extends from boot to bocal, and is shorter and narrower than the long joint.

Butt joint (or boot joint): The boot is at the bottom of the instrument and fold over on itself. The boot is attached to the wing joint and long joint. The hand rest/crutch, which can stabilize the right hand, mounts to the boot joint.

Long joint (or bass joint): The long joint connects the bell and the boot and is the longest piece of tubing, and is parallel to the wing joint.

Bell: The bell, with an ornamental rim of ivory or plastic, extends upward.

Hand rest (or crutch): An adjustable comma-shaped apparatus which mounts to the boot joint is used to stabilize bassoonist's right hand.

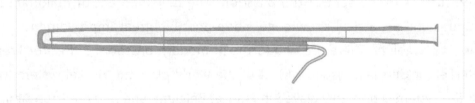

Bore: The bore[4] of a bassoon is conical. The diameter at the very tip of the bocal is around 4 millimeters, while the diameter at the bell it is 40 millimeters. Between those two points, the bore gradually becomes wider, and the walls of the bassoon are thicker at various points along the bore. The two adjoining bores of the boot joint are connected at the bottom of the instrument with a U-shaped metal connector.

Tone hole: The tone holes are drilled at an angle to the axis of the bore, which reduces the distance between the holes on the exterior. Playing becomes easier by closing the distance between the widely spaced holes with a complex system of the keywork, which extends throughout the entire length of the instrument.

II. Development

Origin

The dulcian is the direct forerunner of the modern bassoon, as the two instruments share many characteristics: a double reed fitted to a metal crook, obliquely drilled tone holes and a conical bore that doubles back on itself. The origins of the dulcian are obscure, but by the mid-16th century it was available in as many as eight different sizes, from soprano to great bass. The primary function of the dulcian was used to reinforce the bass line in wind ensembles. Otherwise, dulcian technique was rather primitive, with eight finger holes and two keys, indicating that it could play in only a limited number of key signatures.

The baroque bassoon was a newly invented instrument instead of a simply modification of the old dulcian. In the 1650s, the true bassoon was invented by Martin Hotteterre who created the bassoon in four sections (bell, bass joint, boot and wing joint), an arrangement that allowed greater accuracy in machining the bore compared to the one-piece dulcian. He also extended the compass down to B♭ by adding two keys. Around 1700, a fourth key (G#) was added, and a fifth key, for the low E♭, was added during the first half of the 18th century.

Modern bassoon

Increasing demands on capabilities of instruments and players in the 19th century spurred further refinement of the bassoon. The development of manufacturing techniques and acoustical knowledge made possible great improvements in the bassoon's playability. There are two forms of modern bassoon: Buffet (or French) and Heckel (or German) systems. Most of the world plays the Heckel system, while the Buffet system is primarily played in France, Belgium, and parts of Latin America. Owing to the ubiquity of the Heckel system in English-speaking countries, the contemporary bassoon always means the Heckel system, with the Buffet system being explicitly qualified where it appears.

Heckel (or German) system

In about 1825, the performer, teacher, and composer Carl Almenraeder from Germany, developed a 17-key modern bassoon with a range spanning four octaves.

In 1831, Almenräder started his own factory with a partner, Johann Adam Heckel[5]. Heckel and two generations of descendants continued to refine the bassoon, and because of their superior singing tone quality, the Heckel-style German model of bassoons with its 24–27 keys and five open finger holes, have dominated the field and became the international standard since the 20th century.

Buffet (or French) system

The Buffet system bassoon made an earlier acoustical achievement than Heckel, but it developed in a more conservative way. The modern Buffet system has 22 keys and six open finger holes with its range being the same as the Heckel. Compared to the Heckel bassoon, Buffet system bassoons have a narrower bore and simpler mechanism, requiring different, and often more complex fingerings for many notes. The Buffet system bassoon needs improving consistency of intonation, ease of operation and increasing power, which are found in Heckel bassoon, but the Buffet is considered by some to have a more vocal and expressive quality, and is known for its superior tonal quality in the upper register. Switching between Heckel and Buffet requires extensive retraining.

Contrabassoon[6]

The first useful contrabassoon (or double bassoon), sounding an octave lower than the bassoon and much employed in large scores, was developed in Vienna and used occasionally by the classical composers. The modern contrabassoon follows Heckel's design of approximately 1870, with the tubing doubled back four times and often with a metal bell that points downward.

III. Performing Techniques

In performance, the bassoonist holds the instrument diagonally across the body on a shoulder harness[7] (when standing) or a seat strap[8] (when sitting) due to its huge size and heavy weight.

Embouchure and sound production

The proper bassoon embouchure is a very important aspect of producing a full, round, and rich sound on the instrument. The lips are both rolled over the teeth, often with the upper lip further along in an "overbite". The lips provide micromuscular pressure on the entire circumference of the reed, which grossly controls intonation and harmonic excitement, and thus must be constantly modulated with every change of note. The sound is produced by rolling both lips over the reed and blowing direct air pressure to cause the reed to vibrate.

Modern fingering

The fingering system of the bassoon is quite complex when compared to those of other instruments. The fingering technique of the bassoon varies more between players, and even the technique differs on a global scale. The bassoon is played with both hands in a stationary position, the left above the right, with the five main finger holes on the front of the instrument (nearest the audience) plus a sixth that is activated by an open-standing key[9]. Five additional keys on the front are controlled by the little fingers of each hand. The back of the instrument (nearest the player) has twelve or more keys to be controlled by the thumbs. The left thumb operates nine or ten keys, while the four fingers (the index finger, the middle finger, the ring finger, and the smallest finger) of the left hand can each be used in two different positions. The right thumb operates four keys, while the four fingers of the right hand have at least one assignment each.

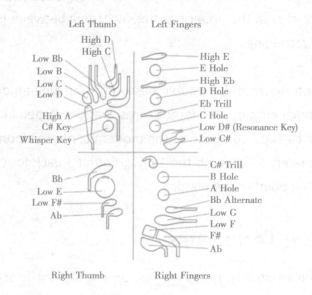

diagram describing the keys on a bassoon

Extended techniques

Many extended techniques can be performed on the bassoon, such as flutter tonguing, staccato, single tonguing, double tonguing, triple tonguing, vibrato and glissando.

IV. Musical Classics

Bassoon Concerto in C Major, FWV L: C2   Johann Friedrich Fasch
（《C 大调巴松管协奏曲》 L:C2 约翰·弗里德里希·法施）

Clarinet & Bassoon Concertos　　Carl Maria von Weber

　　　　（《F 大调巴松管协奏曲》Op. 75 卡尔·马利亚·冯·韦伯）

Bassoon Concerto in A Minor, RV 498　　Antonio Vivaldi

　　　　（《A 小调低音管协奏曲》, RV498 安东尼奥·维瓦尔第）

Concerto for Bassoon in E Minor RV484　　Antonio Vivaldi

　　　　（《E 小调巴松管协奏曲》RV484 安东尼奥·维瓦尔第）

Bassoon Concerto No.1 in B-Flat Major — "La Notte", RV 501　　Antonio Vivaldi

　　　　（《第一降 B 大调巴松管协奏曲"夜晚"》RV 501 安东尼奥·维瓦尔第）

Bassoon Concerto In B Flat, K 191　　Wolfgang Amadeus Mozart

　　　　（《降 B 大调巴松管协奏曲》K 191 沃尔夫冈·阿玛多伊斯·莫扎特）

Bassoon Concerto in E-Flat Major, W. C82　　Johann Sebastian Bach

　　　　（《降 E 大调巴松管协奏曲》W. C82 约翰·塞巴斯蒂安·巴赫）

Bassoon Concerto in B-Flat Major, W. C83　　Johann Sebastian Bach

　　　　（《降 E 大调巴松管协奏曲》W. C83 约翰·塞巴斯蒂安·巴赫）

Grand Concerto for Bassoon and Orchestra in F　　Johann Nepomuk Hummel

　　　　（《F 大调巴松管协奏曲》WoO 23, S63 约翰·尼波默克·胡梅尔）

Bassoon Concerto　　André Jolivet　　　　（《巴松管协奏曲》安德烈·若利韦）

The Five Sacred Trees　　John Williams

　　　　（《五颗圣树(大管与管弦乐队协奏曲)》约翰·威廉姆斯）

Concerto in C Major for Bassoon　　Johann Baptist Georg Neruda

　　　　（《C 大调大管协奏曲》约翰·巴普蒂斯特·格奥尔格·聂鲁达）

Bassoon Concerto in C Major　　Jan Antonín Koželuh

　　　　（《C 大调巴松管协奏曲》利奥波德·科策卢）

Bassoon Concerto　　Jean Francaix　　　　（《巴松管协奏曲》让·弗朗赛）

Bassoon Concerto in B-Flat Major, C73　　Antonio Rosetti

　　　　（《降 B 大调巴松管协奏曲》C73 安东尼·罗赛蒂）

Peter and the Wolf　　Sergei Prokofiev　　（《彼得与狼》谢尔盖·普罗科菲耶夫）

Notes on the text

1. Buffet (or French) system 法式系统
2. Heckel (or German) system 黑格尔（德式）系统
3. conical tube 圆锥管
4. bore 内径，孔径
5. Johann Adam Heckel 约翰·亚当·黑克尔 (1812—1877)，德国乐器制作师。

6. contrabassoon (or double bassoon) 倍低音大管
7. shoulder harness 肩带
8. seat strap 坐带
9. open-standing key 弱音键

Terms

double reed	/ˈdʌbl riːd/	双簧
bassoonist	/bəˈsuːnɪst/	大管演奏者；巴松管手
contrabassoon/ double bassoon	/ˈkɑːntrəbəsuːn/, /ˈdʌbl bəˈsuːn/	低音大管
conical tube	/ˈkɑːnɪkl tuːb/	圆锥管
bore	/bɔːr/	内径，孔径
bell	/bel/	喇叭口
bell joint	/bel dʒɔɪnt/	上节管
crook (or bocal)	/krʊk/	S 管
bass joint / long joint	/beɪs dʒɔɪnt/, /lɔːŋ dʒɔɪnt/	长管
tenor joint / wing joint	/ˈtenər dʒɔɪnt/, /wɪŋ dʒɔɪnt/	支管
strap ring	/stræp rɪŋ/	平衡杆；背带环
hand rest/ crutch	/hænd rest/, /krʌtʃ/	手托
double joint / boot joint/ butt joint	/ˈdʌbl dʒɔɪnt/, /buːt dʒɔɪnt/, /bʌt dʒɔɪnt/	底管
protector cap (contains the U-tube)	/prəˈtektər kæp/	金属帽（含U形管）
embouchure	/ˌɑːmbʊˈʃʊr/	口型
overblowing	/ˌoʊvərˈbloʊɪŋ/	吹出泛音
single tonguing	/ˈsɪŋgl ˈtʌŋɪŋ/	单吐
vibrato	/vɪˈbrɑːtoʊ/	颤音
double tonguing	/ˈdʌbl ˈtʌŋɪŋ/	双吐
triple tonguing	/ˈtrɪpl ˈtʌŋɪŋ/	三吐
trill	/trɪl/	颤音
glissando	/glɪˈsændoʊ/	滑音
staccato	/stəˈkɑːtoʊ/	断音；断奏

Exercises

I. Comprehension questions

1. Which one is the largest member of the woodwind family?
2. What was the primary function of the dulcian in wind ensembles?
3. What is the structure of the bore of the bassoon? Describe it in detail.
4. What is the only performing technique for the bassoon?
5. What are the main differences between Buffet system basson and Heckel system basson?

II. Translating useful expressions

1. 恰当的吹口是巴松发出丰满、圆润、浑厚声音的一个非常重要的因素。
2. 巴松是低音、双簧木管乐器，音乐主要在低音谱号和次中音谱号记谱。
3. 和其他乐器相比，巴松的指法系统更复杂。
4. The bore of a bassoon is conical. The diameter at the very tip of the bocal is around 4 millimeters, while the diameter at the bell it is 40 millimeters.
5. The modern contrabassoon follows Heckel's design, with the tubing doubled back four times and often with a metal bell that points downward.
6. Many performing techniques can be performed on the bassoon, such as flutter tonguing, staccato, single tonguing, double tonguing, triple tonguing, vibrato, trill, and glissando.

III. Brainstorm

The bassoon, as a bass instrument in the woodwind family, is indispensable in an ensemble, but it is hard to be a protagonist (主角). Most people are ordinary and just a supporting role in the world. From this aspect, brainstorm what we can learn from the bassoon and how you understand a protagonist and a supporting role (配角) in your life.

Lesson Nine
Percussion Instruments

Lesson Nine
Percussion Instruments

Percussion instruments are believed to be one of the oldest in the world because they are the simplest to make and play. The first drums are believed to have existed before 6000 BC. The term *percussion* derives from the Latin verb *percussio* to beat, strike in the musical sense, and the noun *percussus*, a beating.

By definition, a percussion instrument is a musical instrument that is sounded by being struck or scraped[1] by a beater; struck, scraped or rubbed[2] by hand; or struck against another instrument(s). It is commonly referred to as "the backbone" or "the heartbeat" of a musical ensemble, often working in close collaboration with bass instruments, when present.

However, most classical pieces written for a full orchestra since the time of Haydn and Mozart are orchestrated to emphasize the strings, woodwinds, and brass. Percussion instruments just served to provide additional accents when needed and were rarely played continuously. In the 18th and 19th centuries, other percussion instruments (like the triangle or cymbals) have been used, again generally sparingly, whereas in the twentieth-century classical music, percussion instruments can be found more frequently.

Today, the most common percussion section seen in an orchestra includes instruments such as timpani, snare drum, bass drum, cymbals, triangle, and tambourine. It can also contain non-percussive instruments, such as whistles[3] and sirens[4], or a blown conch shell[5]. On the other hand, keyboard instruments, such as the marimba, are not typically part of the percussion section, but the glockenspiel and xylophone are included. In jazz and other popular music ensembles, there is normally a drum set with a "crazy" drummer sitting behind.

I. Classification

There are various criteria for classifying percussion instruments, including their construction, origin, methods of sound production, or musical function. The two common classifications are as follows.

By methods of sound production

According to The Hornbostel–Sachs system[6], most percussion instruments are classified as idiophones[7] and membranophones[8].

Idiophones are instruments whose own substance vibrates to produce sound, such as bells, clappers, rattles, and so on. Membranophones are sounded by the vibration of a stretched membrane, including all types of drums. It is important to note that not all instruments in these groups are sounded by being struck. Other playing methods include rubbing, shaking, plucking, and scraping. Although many idiophones and some membranophones are tunable and hence may be melody instruments, both groups typically serve to delineate or emphasize rhythm. Percussion instruments form the third section of the modern Western orchestra, with stringed and wind instruments making up the other two sections.

By musical function or orchestration

Percussion instruments can be categorized as pitched percussion instruments, which produce sounds with an identifiable pitch, and unpitched percussion instruments, which produce notes with an unidentifiable pitch.

As their name suggests, pitched percussion instruments can produce identifiable musical tones belonging to one or more pitches. For example, the xylophone has distinct sounds produced by each bar or key, regardless of how hard it is struck. Other pitched percussion instruments include the steelpans and the tubular bells.

On the other hand, unpitched percussion instruments have no definite pitches. They are often used to provide rhythm or accents and are played independently of a song's harmony and melody. Some examples are the snare drum, cymbals, tambourine, and triangle.

Interestingly, some percussion instruments can fall under both categories, depending on how they are used. Bells, for example, can be pitched or unpitched. Knowing the category to which an instrument belongs makes it easier to determine how to maximize its potential for the desired piece.

II. Major Percussion Instruments

Timpani

The timpani is also called kettle drum owing to their distinctive size and shape. Typically the body of the large "kettle", made of brass or copper, provides a very resonant and booming sound. The head[9], historically of stretched animal skin, is now commonly made of plastic or a material called "vellum"[10]. In the classical period, a set of timpani usually consisted of two drums, tuned to each the tonic and dominant of a piece, and could not be retuned quickly and accurately due to a lack of mechanical innovation. Today, the timpani's head is mechanically controlled by a foot pedal, which allows for dynamic adjustment of tension. This provides the timpanist with a wide range of tuning options that can be quickly and accurately adjusted after practice. To quickly determine the desired pitch, a small pitch-pipe[11] can be used. Another innovative option is the mechanical dial[12] which indicates the note the timpanist will produce under the given tension, which provides for relatively significant changes in tuning in a short period with sufficient practice.

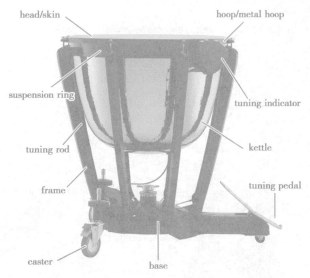

Snare drum

The snare drum or side drum is a ubiquitous percussion instrument known for its cylindrical shape and powerful, staccato[13] sound. It originated from the Tabor drum[14], which was initially used to accompany the flute, and then has evolved into various modern versions including the kit snare, marching snare, the piccolo snare[15], etc. Each of them presents a different dimension and style. Incidentally, the snare drum became an indispensable component of the drum set (see also 4) *Drum set*) in the late 19th century.

The snare drum finds frequent usage in orchestras, concert bands, marching bands, parades, drum corps, and more. It is typically played with drum sticks, although there are other options for a completely different sound, such as brushes, mallets, or even bare hands.

The snare drum has two heads—both usually made of plastic—along with a rattle of metal wires on the bottom head called the snares[16]. The top head is called the batter head[17] because that is where the drummer strikes it, while the bottom head is called the snare head[18] because that is where the snares are located. The tension of the drum heads is held constant through the adjustable tension rods[19] which provide an opportunity to differ the sound of the hit. The strainer[20] is a lever that controls the release and tightness of the snares. When the strainer is relaxed, the sound of the snare is more like that of a tom-tom. The rim[21], a metal ring around the batter head, serves mutiple purposes, including the production of a piercing rimshot when struck with a drumstick.

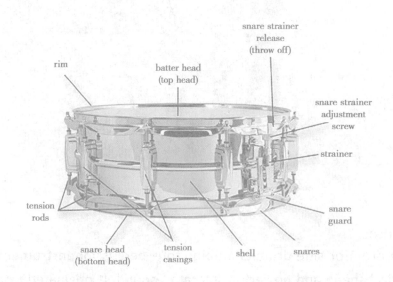

Parts of the Snare Drum

Marimba

The marimba is a percussion instrument consisting of a set of wooden bars that are struck with mallets to produce musical tones. Resonators attached to the bars amplify their sound. The bars are arranged as those of a piano, with the accidentals raised vertically and overlapping the natural bars, aiding the performer both visually and physically. This instrument belongs to the idiophone family, but possesses a more resonant and lower-pitched tessitura compared to the xylophone.

The chromatic marimba was developed in Chiapas, Mexico, evolving from the local diatonic marimba[22], an instrument whose ancestor was a type of balafon[23] built by African slaves in Central America.

Modern uses of the marimba include solo performances, woodwind and brass ensembles, marimba concertos, jazz ensembles, marching band (front ensembles), drum and bugle corps, and orchestral compositions. Contemporary composers have used the unique sound of the marimba more and more in recent years.

One key element contributing to the marimba's rich sound is its resonators. These are tubes (usually made of aluminum) that hang below each bar. In the most traditional versions, various sizes of natural gourds are attached below the keys to act as resonators; in more sophisticated versions, carved wooden resonators are used, allowing for more precise tuning of the pitch. In Central America and Mexico, a hole is often carved into the bottom of each resonator and then covered with a delicate membrane taken from a pig's intestine, resulting in a characteristic "buzzing" or "rattling" sound known as charleo[24]. In more contemporary-style marimbas, wood is replaced by PVC tubing, and the holes in the bottoms of the tubes are covered with a thin layer of paper to produce the buzzing noise.

The length of the resonators varies based on the frequency produced by each bar. Vibrations from the bars resonate as they pass through the tubes, amplifying the tone in a manner very similar to how the body of a guitar or cello would.

Drum set

A drum set—also called a drum kit, trap set or simply drums—is a collection of drums, cymbals and other percussion instruments, which are set up on stands to be played by a single musician. The player holds drumsticks in both hands and operates pedals with their feet to control the hi-hat cymbal and the bass drum beater. A drum kit consists of a mix of drums (categorized classically as membranophones, Hornbostel-Sachs high-level classification) and idiophones—primarily cymbals, but may also include the woodblock and cowbell (classified as Hornbostel-Sachs high-level classification 1). In the 2020s, some kits also incoporate electronic instruments

(Hornbostel-Sachs classification 53).

A standard modern kit (for a right-handed player[25]), used in popular music and taught in music schools, contains:

- A snare drum, mounted on a stand, placed between the player's knees and played with drumsticks or brushes;
- A bass drum, played by a pedal operated by the right foot, which moves a beater;
- Two or more toms, played with sticks or brushes;
- A hi-hat (two cymbals mounted on a stand), played with sticks or brushes, opened and closed with left foot pedal (it can also produce sound with the foot alone);
- One or more cymbals, mounted on stands, played with sticks or brushes.

Many drummers expand upon this basic setup by adding additional drums, cymbals, and other instruments including pitched percussion. In certain music styles, specific additions are common. For example, some rock and heavy metal drummers make use of double bass drums, achieved with either a second bass drum or a remote double foot pedal[26]. Some progressive drummers may include orchestral percussion such as gongs and tubular bells in their rig. Some performers, such as some rockabilly[27] and funk[28] drummers, play smaller kits that omit elements from the basic setup.

The drum kit is a part of the standard rhythm section, used in many types of popular and traditional music styles, ranging from rock and pop to blues and jazz. Other common instruments used in the rhythm section include the piano, electric guitar, electric bass, and keyboards.

Beater

There are several names for the tools wielded by percussionists — beater, drum stick and mallet, for instance. The term beater is slightly more general, for it may sometimes refer to a foot or mechanically operated device, e.g. *bass-drum pedal*. The word *drumstick* is more specific but still applied to a wide range of beaters. A mallet is normally held in the hand. Some mallets, such as a triangle beater, are used only with a specific instrument, while others are used across many different instruments. Often, mallets of different materials and hardness are used to achieve various timbres on the same types of instrument (e.g. using either wooden or yarn mallets on a marimba).

Mallet

Mallet is a term used in percussion playing for any stick, beater or hammer that has a head, whether it be the small felt ball for the timpani stick, or the large weighted head needed to bring out the tone of a tam-tam. Keyboard percussion instruments such as the vibraphone, marimba and xylophone are categorized as "mallet instruments". There are three main types:

- *Unwrapped mallets, used on glockenspiel, xylophone, and other instruments with keys made of durable material, have heads made of brass, rubber, nylon, wood, or other hard materials.*
- *Wrapped mallets, mostly used on marimba, vibraphone, and other instruments with softer keys (though they can be used on more durable instruments as well), have heads of rubber, nylon, acrylic[29] or other medium-hard materials that are wrapped in softer materials like yarn, cord or latex. Wrapped mallets are also a choice for playing suspended cymbal, though drum set players will typically use drum sticks instead.*
- *Felt mallets or cartwheel mallets[30] have heads composed of layers of felt, held between two steel washers. They are mainly used on untuned percussion as well as on timpani.*

Mallet shafts are commonly made of rattan, birch, or synthetic materials such as fibreglass. Birch is stiff and typically longer, while rattan shafts offer more flexibility and a more open sound. Fiberglass is ideal for playing lightly on an instrument because it is easy to control.

Drumstick

A drumstick is a type of percussion mallet used particularly for playing snare drum, drum kit, and some other percussion instruments, especially those that produce unpitched sounds.

Specialized beaters used on some other percussion instruments, such as the metal beater used for a triangle or the mallets used with tuned percussion (like xylophone and timpani), are not normally referred to as drumsticks.

Drumsticks come in many different types, each suited for various instruments:

- Rutes, used with a wide range of instruments.
- Brushes, used particularly with snare drum but also with many other instruments.
- Bachi[31], used with Japanese taiko drums[32].

Notes on the text

1. scraped 刮擦（演奏方式）
2. rubbed 揉搓（演奏方式）
3. whistles 口哨
4. sirens 汽笛
5. conch shell 海螺壳
6. The Hornbostel–Sachs system 霍恩波-斯特尔与萨克斯乐器分类法
7. idiophones 体鸣乐器
8. membranophones 膜鸣乐器
9. head 鼓面；鼓皮
10. vellum 仿制皮纸
11. pitch-pipe 定音笛
12. mechanical dial 调音指示器
13. staccato 此处指军鼓音色短促、刺耳
14. Tabor drum 塔波鼓
15. piccolo snare 高音军鼓
16. snares 军鼓响弦
17. batter head 打击面
18. snare head 共鸣面
19. tension rods 张力螺丝
20. strainer 军鼓上弦器
21. rim 压圈
22. diatonic marimba 全音马林巴
23. balafon 巴罗琴
24. charleo 嚓哩欧
25. a right-handed player 现代架子鼓默认为惯用手为右手的鼓手设计
26. double foot pedal 双踩，鼓手常用拓展设备之一
27. rockabilly 乡村摇滚乐，音乐风格
28. funk 放克音乐，音乐风格
29. rubber, nylon, acrylic 橡胶、尼龙、丙烯酸塑料，制作鼓皮的常见材料
30. cartwheel mallets 一种槌头形状像车轮一样的鼓槌
31. Bachi 太鼓槌
32. taiko drums 太鼓

Lesson Nine
Percussion Instruments

🎻 Terms

1) Common percussion instruments

head/skin	/hed/, /skɪn/	鼓皮
suspension ring	/səˈspenʃn rɪŋ/	悬挂环
tuning rod	/ˈtuːnɪŋ rɑːd/	调音杆
frame	/freɪm/	鼓架
caster	/ˈkæstər/	脚轮
base	/beɪs/	底盘
tuning pedal	/ˈtuːnɪŋ ˈpedl/	调音踏板
kettle	/ˈketl/	鼓身；鼓桶
tuning indicator	/ˈtuːnɪŋ ˈɪndɪkeɪtər/	调音指示器
hoop/metal hoop	/huːp/, /ˈmetl huːp/	鼓圈
batter head/top head	/ˈbætər hed/, /tɑːp hed/	鼓皮；打击面
rim	/rɪm/	（鼓）压圈
tension rod	/ˈtenʃn rɑːd/	张力螺丝
snare head/bottom head	/sner hed/, /ˈbɑːtəm hed/	共鸣面
tension casing/lug	/ˈtenʃn ˈkeɪsɪŋ/, /lʌg/	鼓耳
shell	/ʃel/	鼓腔
snare	/sner/	响弦
snare guard	/sner gɑːrd/	响弦固定器
strainer	/ˈstreɪnər/	上弦器；响弦器
bass drum	/beɪs drʌm/	大鼓
xylophone	/ˈzaɪləfəʊn/	木琴
tom tom	/tɑːm tɑːm/	嗵嗵鼓
triangle	/ˈtraɪæŋgl/	三角铁
glockenspiel	/ˈglɑːkənʃpiːl/	钟琴
vibraphone	/ˈvaɪbrəfəʊn/	颤音琴
gong	/gɔːŋ/	大锣
conga	/ˈkɑːŋgə/	康加鼓
bongo	/ˈbɑːŋgəʊ/	邦戈鼓
pair cymbals	/per ˈsɪmbəlz/	对镲
suspended cymbal	/səˈspendɪd ˈsɪmbəl/	吊镲
hi-hat	/ˈhaɪ hæt/	踩镲
ride cymbal	/raɪd ˈsɪmbəl/	叮叮镲
tambourine	/ˌtæmbəˈriːn/	铃鼓
temple block	/ˈtempl blɑːk/	木鱼
marching snare	/ˈmɑːrtʃɪŋ sner/	行进军鼓
brush	/ˈbrʌʃ/	鼓刷

2) Techniques

single stroke	/ˈsɪŋgl strəʊk/	单击
double stroke	/ˈdʌbl strəʊk/	双跳
paradiddle	/ˌpærəˈdɪdəl/	复合跳
open roll	/ˈoʊpən roʊl/	开放型滚奏
matched grip	/mætʃt grɪp/	扣腕握槌法
closed roll	/kləʊzd roʊl/	密集型滚奏
rimshot	/ˈrɪm ʃɑːt/	鼓边平击；重音边击
cross-stick	/ˈkrɔːs stɪk/	止音边击
dead stroke	/ded strəʊk/	压奏
traditional grip	/trəˈdɪʃənl grɪp/	传统握槌法
Burton grip	/ˈbɜːrtn grɪp/	巴顿握槌法
Musser grip	/myse grɪp/	穆瑟握槌法
Stevens grip	/ˈstivənz grɪp/	斯蒂文斯握槌法

Exercises

I. Comprehension questions

1. What are the normal classifications according to this section?
2. What are the two types that the Hornbostel–Sachs system divides the most percussion instruments into?
3. Why does the timpani get the name "kettle drum"?
4. Where is the snare drum often used?
5. Which factor determines the length of the resonators?
6. How many snare drums are included in a standard modern drum kit?

II. Translating useful expressions

1. 打击乐器被认为是存在历史最长的乐器。
2. 定音鼓的主要结构包括鼓皮、鼓身以及调音系统。
3. 小军鼓的鼓腔材质非常多样，如铜质、铝质、枫木以及桦木等。
4. 马林巴的演奏十分困难，需要两只手分别控制两支琴槌进行独立地运动。
5. 架子鼓可以用于演奏多种音乐风格，比如爵士、布鲁斯和摇滚。

6. Today, the most common percussion section to be seen in an orchestra includes instruments such as timpani, snare drum, bass drum, cymbals, triangle, and tambourine.
7. There are various criteria for classifying percussion instruments, such as their construction, origin, methods of sound production, or musical function.
8. In the classical period, a set of timpani usually consisted of two drums, tuned to each the tonic and dominant of a piece, and could not be retuned quickly and accurately due to lack of mechanical innovation.
9. The marimba is a percussion instrument consisting of a set of wooden bars (音板) struck with mallets to produce musical tones
10. Many drummers extend their kits from this basic configuration, adding more drums, more cymbals, and many other instruments including pitched percussion.

III. Brainstorm

With the development of globalization, our world has undergone great changes. Many scholars believe that the drum kit was born in this process. Now, let's consider how this wave of development affects your profession. Is it an opportunity or a challenge?

Lesson Ten

Synthesizer

A synthesizer (also spelled synthesiser) is an electronic musical instrument that generates audio signals. Synthesizers typically create sounds by generating waveforms through methods including subtractive synthesis, additive synthesis and frequency modulation synthesis. These sounds may be altered by components such as filters, which cut or boost frequencies; envelopes, which control articulation, or how notes begin and end; and low-frequency oscillators, which modulate parameters such as pitch, volume, or filter characteristics affecting timbre. Synthesizers are typically played with keyboards or controlled by sequencers, software or other instruments, and may be synchronized to other equipment via MIDI.

I. Development

Precursors

As electricity became more widely available, the early 20th century saw the invention of electronic musical instruments including the Telharmonium[1], Trautonium[2], Ondes Martenot[3], and theremin[4]. In 1957, synthesizer-like instruments emerged in the United States, such as the RCA Mark II, which was controlled with *punch cards*[5] and used hundreds of vacuum tubes.

1960s: Early years

In 1964, the Moog synthesizer debuted. Designed by the American engineer Robert Moog[6], the synthesizer was composed of separate modules which created and shaped sounds, connected by patch cords. Moog developed a means of controlling pitch through voltage, the voltage-controlled oscillator, which along with Moog components such as envelopes, noise generators, filters, and sequencers, became standard components in synthesizers.

1970s: Portability, polyphony and patch memory[7]

In 1970, Moog launched a cheaper, smaller synthesizer, the **Minimoog**. The Minimoog was the first synthesizer sold in music stores, and was more practical for live performance; it standardized the concept of synthesizers as self-contained instruments with built-in keyboards.

Early synthesizers were monophonic, meaning they could only play one note at a time. In 1978, the American company Sequential Circuits[8] released the Prophet-5[9], the first fully programmable polyphonic synthesizer, which used microprocessors to allow users to store sounds in patch memory for the first time, while previous synthesizers required users to adjust cables and knobs to change sounds, with no guarantee of exactly recreating a sound.

1980s: Digital technology

The synthesizer market grew dramatically in the 1980s. 1982 saw the introduction of MIDI, a standardized means of synchronizing electronic instruments, which remains an industry standard[10]. The Yamaha DX7, released in 1983, was the first commercially successful and popularized digital synthesis. Based on frequency modulation (FM) synthesis, the DX7 was characterized by its "harsh", "glassy" and "chilly" sounds, compared to the "warm" and "fuzzy" sounds of analog synthesis.

Digital synthesizers typically contained preset sounds emulating acoustic instruments, with algorithms controlled with menus and buttons. The advent of digital synthesizers led to a downturn in interest in analog synthesizers.

1990s–present: Software synthesizers and analog revival

1997 saw the release of ReBirth, the first software synthesizers that could be played in real time via MIDI. Software synthesizers now can be run as plug-ins or embedded on microchips.

The market for patchable and modular synthesizers rebounded in the late 1990s. In the 2000s, older analog synthesizers regained popularity, sometimes selling for much more than their original prices. In the 2010s, new, affordable analog synthesizers were introduced by companies. The renewed interest is credited to the appeal of imperfect "organic" sounds and simpler interfaces, and modern surface-mount technology[11] making analog synthesizers cheaper and faster to manufacture. Synthesizers were initially viewed as avant-garde, valued by the 1960s psychedelic[12] and counter-cultural[13] scenes but with little perceived commercial potential. Switched-On Bach (1968), a bestselling album of Bach compositions arranged for synthesizer by Wendy Carlos[14], took synthesizers to the mainstream. Synthesizers were adopted by electronic acts and pop and rock groups in the 1960s and 1970s, and were widely used in 1980s music. Today, the synthesizer is used in nearly every genre of music, and is considered one of the most important instruments in the music industry.

II. Sound Synthesis

Synthesizers generate audio through various forms of analogue and digital synthesis.

In subtractive synthesis, complex waveforms are generated by oscillators and then shaped with filters to remove or boost specific frequencies. Subtractive synthesis is characterized as "rich" and "warm".

In additive synthesis, a large number of waveforms, usually *sine waves*, are combined into a composite sound

In frequency modulation (FM) synthesis, also known as phase modulation, a carrier wave is modulated with the frequency of a *modulator wave*; the resulting complex waveform can, in turn, be modulated by another modulator, and this by another, and so on. FM synthesis is characterized as "harsh", "glassy" and "chilly".

In wavetable synthesis, synthesizers modulate smoothly between digital representations[15] of different waveforms, changing the shape and timbre.

In sample-based synthesis, instead of sounds being created by synthesizers, samples (digital recordings of sounds) are played back and shaped with components such as filters, envelopes and LFOs.

III. Components

Oscillators

Oscillators are the fundamental of whole synth, which produce different waveforms (such as sawtooth, sine, square, or pulse waves) with different timbres. By combining different waves, thousands of new sounds are produced. However, oscillators don't have enough ability to create vivid sounds, so filters are needed to enrich its sounds.

Voltage-controlled amplifiers

Voltage-controlled amplifiers (VCAs) control the volume or gain of the audio signal. VCAs can be modulated by other components, such as LFOs and envelopes. A VCA is a preamp that boosts (amplifies) the electronic signal before passing it on to an external or built-in power amplifier, as well as a means to control its amplitude (volume) using an attenuator. The gain of the VCA is affected by a control voltage (CV), coming from an envelope generator, an LFO, the keyboard or some other source.

Filters

Voltage-controlled filters (VCFs) "shape" the sound generated by the oscillators in the frequency domain, often under the control of an envelope or LFO. These are essential to subtractive synthesis. Filters are particularly important in subtractive synthesis, being designed to pass some frequency regions (or "bands") through unattenuated while significantly attenuating ("subtracting") others.

The low-pass filter (LPF) keeps lower frequency while high-pass filter (HPF) kicks out the lower frequency we don't need. The band-pass filter (BPF) only keeps frequency of surrounds of the cutoff point. The cutoff point is like a dead-end of frequency, because frequency stepping over the cutoff point will be filtered by filter. The low-pass filter is most frequently used, but band-pass filters, band-reject filters and high-pass filters are also sometimes available.

Envelopes

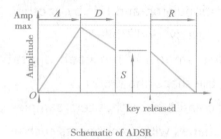

Schematic of ADSR

Envelopes control how sounds change over time. They may control other parameters such as amplitude (volume), filters (frequencies), or pitch, not only the level of volume. The most common envelope is the ADSR (attack, decay, sustain, release) envelope:

Attack is the time taken for initial run-up of level from nil to peak, beginning when the note is triggered.

Decay is the time taken for the subsequent run-down from the attack level to the designated sustain level.

Sustain is the level during the main sequence of the sound's duration, until the key is released.

Release is the time taken for the level to decay from the sustain level to zero after the key is released.

Low-frequency oscillators

Low-frequency oscillators (LFOs) produce waveforms used to modulate parameters, such as the pitch of oscillators (producing vibrato).

Arpeggiators

Arpeggiators, included in many synthesizer models, take input chords and convert them into arpeggios. They usually include controls for speed, range and mode (the movement of the arpeggio).

IV. Controllers

Synthesizers are often controlled with electronic or digital keyboards or MIDI controller keyboards, which may be built into the synthesizer unit or attached via connections such as CV/gate, USB, or MIDI. Keyboards may offer expression such as velocity sensitivity and aftertouch, allowing for more control over the sound.

Other controllers include ribbon controllers, wind controllers, and motion-sensitive controllers etc.

Notes on the text

1. Telharmonium 电传乐器；电传簧风琴
2. Trautonium 特劳托宁电子琴
3. Ondes Martenot 马特诺电子琴；马特诺音波琴
4. theremin 泰勒明电子琴
5. punch cards 穿孔卡：依照固定模式打孔，作为代码，贮存和阅读资料的卡。
6. Robert Moog 罗伯特·穆格 (1934—2005)，被称为"合成器之父"，是电子音乐的先锋。
7. patch memory/ memory patch 内存补丁
8. the American company Sequential Circuits 美国时序电路公司
9. Prophet-5 先知 -5，1978 年和 1984 年之间生产的一款模拟合成器。
10. industry standard 行业标准
11. surface-mount technology 表面贴装技术
12. psychedelic 迷幻摇滚
13. counter-cultural 嬉皮文化；反文化
14. Wendy Carlos 温迪·卡洛斯 (女)，原名沃尔特·卡洛斯 (Walter Carlos/ 男)，是世界上伟大的电子合成器革新者之一、电子音乐的先驱。
15. digital representations 数字表示

Terms

subtractive synthesis	/səbˈtræktɪv ˈsɪnθəsɪs/	减法合成
additive synthesis	/ˈædətɪv ˈsɪnθəsɪs/	加法合成
frequency modulation (FM) synthesis	/ˈfriːkwənsi ˌmɑːdʒəˈleɪʃn ˈsɪnθəsɪs/	调频（FM）合成器
wavetable synthesis	/ˈweɪvteɪbl ˈsɪnθəsɪs/	波表合成
sample-based synthesis	/ˈsæmpl beɪst ˈsɪnθəsɪs/	采样合成
analog synthesis	/ˈænəlɔːg ˈsɪnθəsɪs/	模拟合成器
parameter	/pəˈræmɪtər/	参数
synchronize	/ˈsɪŋkrənaɪz/	同步
sequence	/ˈsiːkwəns/	音序器
modulation	/ˌmɑːdʒəˈleɪʃn/	调制；调节；改变
MIDI: Musical Instrument Digital Interface	/ˈmɪdi/: /ˈmjuːzɪkl ˈɪnstrəmənt ˈdɪdʒɪtl ˈɪntərfeɪs/	电子音乐数字接口
algorithm	/ˈælgərɪðəm/	算法

plug-in	/ˈplʌg ɪn/	插件
carrier wave	/ˈkæriər weɪv/	载波
oscillator (OSC)	/ˈɑːsɪleɪtər/	振荡器
waveform	/ˈweɪvfɔːrm/	波形
sawtooth wave	/ˈsɔːˌtuːθ weɪv/	锯齿波
sine wave	/saɪn weɪv/	正玄波
square wave	/skwer weɪv/	方波
pulse wave	/pʌls weɪv/	脉冲波
preamp	/ˈpriæmp/	前置放大器
amplitude	/ˈæmplɪtuːd/	幅度
control voltage	/kənˈtroʊl ˈvoʊltɪdʒ/	控制电压
attenuator	/əˈtenjueɪtər/	消音器；衰减器
envelope	/ˈɑːnvələʊp/	包络
envelope genetator	/ˈɑːnvələʊp ˈdʒenəreɪtər/	包络发生器
filter	/ˈfɪltər/	滤波器
cutoff point	/ˈkʌt ɔːf pɔɪnt/	截止点
cutoff frequency	/ˈkʌt ɔːf ˈfriːkwənsi/	截止频率
low-pass filter	/loʊ pæs ˈfɪltər/	低通滤波器
high-pass filter	/haɪ pæs ˈfɪltər/	高通滤波器
band-pass filter	/bænd pæs ˈfɪltər/	带通滤波器
attack	/əˈtæk/	起音
decay	/dɪˈkeɪ/	衰减
sustain	/səˈsteɪn/	保持
release	/rɪˈliːs/	释放；尾音
low-frequency oscillator (LFO)	/loʊ ˈfriːkwənsi ˈɑːsɪleɪtər/	低频滤波震荡器
arpeggiator		琶音器
aftertouch	/ˈæftər tʌtʃ/	触后
delay	/dɪˈleɪ/	混响延时；延迟
level/ amp/ volume	/ˈlevl/, /æmp/, /ˈvɑːljəm/	音量（控制单独振荡器的音量）
pitch	/pɪtʃ/	音高调节
velocity	/vəˈlɑːsəti/	力度
duration	/duˈreɪʃn/	时值
tempo	/ˈtempəʊ/	演奏速度
transport	/ˈtrænspɔːrt/	传送；输送
sample	/ˈsæmpl/	采样；样本

Exercises

I. Comprehension questions

1. What musical instrument is a synthesizer?
2. What are the main components in a synthesizer?
3. How do synthesizers create sounds?
4. Who took synthesizers to the mainstream?
5. What is Moog's contribution to the synthesizer?

II. Translating useful expressions

1. 数字合成器通过各种形式的模拟合成和数字合成产生声音。
2. 但是，振荡器无法创造有生命力的声音。
3. 把包络连接到截止点上是使声音活泼的捷径。
4. Synthesizers are typically played with keyboards or controlled by sequencers, software or other instruments, and may be synchronized to other equipment via MIDI.
5. Oscillators are the fundamental of whole synth, which produce different waveforms with different timbres.
6. The low-pass filter (LPF) keeps lower frequency while high-pass filter (HPF) kicks out the lower frequency we don't need.

III. Brainstorm

Today, the synthesizer is used in nearly every genre of music, and is considered one of the most important instruments in the music industry. Please take the synthesizer as an example to research and discuss the differences between the acoustic musical instruments and electronic musical instruments.

Appendix I Singing (Vocal)

Singing is the act of producing musical sounds with the voice and augments regular speech by the use of sustained tonality, rhythm, and a variety of vocal techniques. A person who sings is called a singer or vocalist (in jazz and popular music). Singers perform music (arias, recitatives, songs, etc.) that can be sung with or without accompaniment by musical instruments. Different singing styles include art music such as opera and Chinese opera, Indian music and religious music styles such as gospel, traditional music styles, world music, jazz, blues, ghazal and popular music styles such as pop, rock, electronic dance music and filmi (Indian film songs).

Singing can be formal or informal, arranged or improvised. It may be done as a form of religious devotion, as a hobby, as a source of pleasure, comfort or ritual, as part of music education or as a profession. Excellence in singing requires time, dedication, instruction and regular practice. If practice is done on a regular basis then the sounds can become more clear and strong. Professional singers usually build their careers around one specific musical genre, such as classical or rock, although there are singers with crossover success (singing in more than one genre). Professional singers usually take voice training provided by voice teachers or vocal coaches throughout their careers.

I. Voices

Anatomy and physiology as it relates to the physical process of singing
In its physical aspect, singing has a well-defined technique that depends on the use of the lungs, which act as an air supply or bellows; on the larynx, which acts as a reed or vibrator; on the chest, head cavities and skeleton, which have the function of an amplifier, as the tube in a wind instrument; and on the tongue, which together with the palate, teeth, and lips articulate and impose consonants and vowels on the amplified sound. Though these four mechanisms function independently, they are nevertheless coordinated in the establishment of a vocal technique and are

made to interact upon one another. During passive breathing, air is inhaled with the diaphragm while exhalation occurs without any effort. The pitch is altered with the vocal cords. With the lips closed, this is called humming.

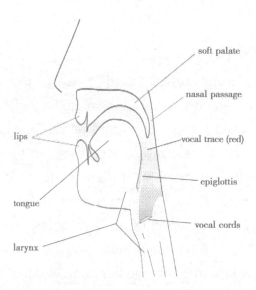

The sound of each individual's singing voice is entirely unique not only because of the actual shape and size of an individual's vocal cords, but also due to the size and shape of the rest of that person's body. Humans have vocal folds which can loosen, tighten, or change their thickness, and over which breath can be transferred at varying pressures. The shape of the chest and neck, the position of the tongue, and the tightness of otherwise unrelated muscles can be altered. Any one of these actions results in a change in voices pitch, volume (loudness), timbre, or tone of the sound produced. Sound also resonates within different parts of the body and an individual's size and bone structure can affect the sound produced by an individual. Singers can also learn to project sound in certain ways so that it resonates better within their vocal tract. This is known as vocal resonation. Another major influence on vocal sound and production is the function of the larynx which people can manipulate in different ways to produce different sounds. These different kinds of laryngeal function are described as different kinds of vocal registers. It has also been shown that a more powerful voice may be achieved with a fatter and fluid-like vocal fold mucosa. The more pliable the mucosa is, the more efficient the transfer of energy from the airflow to the vocal folds becomes.

Vocal registration

Vocal registration refers to the system of vocal registers within the voice. A register in the voice is a particular series of tones, produced in the same vibratory pattern of the vocal folds, and possessing the same quality, which originate in laryngeal function, because each of these vibratory patterns appears within a particular range of pitches and produces certain characteristic sounds. The term "register" can be somewhat confusing as it encompasses several aspects of the voice. The term register can be used to refer to any of the following:

1. A particular part of the vocal range such as the upper, middle, or lower registers.
2. A resonance area such as chest voice or head voice.
3. A phonatory process (phonation is the process of producing vocal sound by the vibration of the vocal folds that is in turn modified by the resonance of the vocal tract)
4. A certain vocal timbre or vocal "color".
5. A region of the voice which is defined or delimited by vocal breaks[1].

There are four vocal registers based on the physiology of laryngeal function from the lowest to the highest: the vocal fry register, the modal register (chest voice), the falsetto register (head voice), and the whistle register.

Vocal resonation

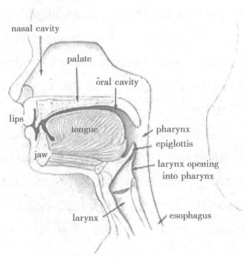

cross-section of the head and neck

Vocal resonation is the process by which the basic product of phonation is enhanced in timbre and/or intensity by the air-filled cavities through which it passes on its way to the outside air. The end result of resonation is, or should be, to make a better sound. There are seven areas that may be listed as possible vocal resonators. In sequence from the lowest within the body to the highest, these areas are the chest, the tracheal tree, the larynx itself, the pharynx, the oral cavity, the nasal cavity, and the sinuses.

Chest voice and head voice

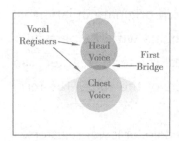

Chest voice and head voice are terms used within vocal music. Chest voice can be used in relation to a particular part of the vocal range or type of vocal register; a vocal resonance area; or a specific vocal timbre. Head voice can be used in relation to a particular part of the vocal range or type of vocal register or a vocal resonance area.

In Men, the head voice is commonly referred to as the falsetto.

The first recorded mention of the terms chest voice and head voice was around the 13th century, but, as knowledge of physiology has increased over the past several hundred years, the understanding of the physical process of singing and vocal production has been improving. As a result, many vocal pedagogists have redefined or even abandoned the use of the terms chest voice and head voice. In particular, the use of the terms chest register and head register have become controversial. The contemporary use of the term chest voice often refers to a specific kind of vocal coloration or vocal timbre.

Classifying singing voices

Voice classification is the process by which human singing voices are evaluated and are thereby designated into voice types. These qualities include but are not limited to vocal range, vocal weight[2], vocal tessitura, vocal timbre, and vocal transition points such as breaks and lifts within the voice. No system is universally applied or accepted on voice classification.

Within classical music, there are seven different major voice categories. Women are typically divided into three groups: soprano, mezzo-soprano, and contralto. Men are usually divided into four groups: countertenor, tenor, baritone, and bass. Within each of these major categories, there are several sub-categories that identify specific vocal qualities like coloratura facility and vocal weight to differentiate between voices.

Within choral music, singers' voices are divided solely on the basis of vocal range. Choral music most commonly divides vocal parts into high and low voices within each sex (soprano, alto, tenor, and bass).

Within contemporary forms of music (sometimes referred to as contemporary commercial music), singers are classified by the style of music they sing, such as jazz, pop, blues, soul, country, folk, and rock styles. There is currently no authoritative voice classification system within non-classical music. Since contemporary musicians use different vocal techniques, microphones, and are not forced to fit into a specific vocal role, applying such terms as soprano, tenor, baritone, etc. can be misleading or even inaccurate.

II. Vocal Pedagogy

Vocal pedagogy is the study of the teaching of singing. The art and science of vocal pedagogy has a long history that began in Ancient Greece and continues to develop and change today. Professions that practice the art and science of vocal pedagogy include vocal coaches, choral directors, vocal music educators, opera directors, and other teachers of singing.

Vocal pedagogy concepts are a part of developing proper vocal technique. Typical areas of study include the following:

1. Anatomy and physiology as it relates to the physical process of singing.
2. Vocal health and voice disorders related to singing.
3. Breathing and air support for singing.
4. Phonation.
5. Vocal resonation or Voice projection.
6. Vocal registration.
7. Voice classification.
8. Vocal styles: for classical singers, this includes styles ranging from Lieder to opera; for pop singers, styles can include "belted out" a blues ballads; for jazz singers, styles can include Swing ballads and scatting.
9. Techniques used in styles such as sostenuto and legato, range extension, tone quality, vibrato, and coloratura

Vocal techniques

Singing when done with proper vocal technique is an integrated and coordinated act that effectively coordinates the physical processes of singing. There are four physical processes involved in producing vocal sound: respiration, phonation, resonation, and articulation. These processes occur in the following sequence:

1. Breath is taken.
2. Sound is initiated in the larynx.
3. The vocal resonators receive the sound and influence it.
4. The articulators shape the sound into recognizable units.
5. Correct posture for singing.

6. Relaxation.

7. Warm up before singing.

Although these four processes (respiration, phonation, resonance, and articulation) are often considered separately when studied, in actual practice, they merge into one coordinated function. Many vocal problems result from a lack of coordination within this process.

The areas of vocal technique which seem to depend most strongly on the student's ability to coordinate various functions are:

1. Extending the vocal range to its maximum potential.

2. Developing consistent vocal production with a consistent tone quality.

3. Developing flexibility and agility.

4. Achieving a balanced vibrato.

5. A blend of chest and head voice on every note of the range.

Developing the singing voice

Singing is a skill that requires highly developed muscle reflexes. Singing does not require much muscle strength but it does require a high degree of muscle coordination. Individuals can develop their voices further through the careful and systematic practice of both songs and vocal exercises. Vocal exercises have several purposes, including warming up the voice; extending the vocal range; "lining up" the voice horizontally and vertically; and acquiring vocal techniques such as legato, staccato, control of dynamics, rapid figurations, learning to sing wide intervals comfortably, singing trills, singing melismas and correcting vocal faults.

Vocal pedagogists instruct their students to exercise their voices in an intelligent manner. Singers should be thinking constantly about the kind of sound they are making and the kind of sensations they are feeling while they are singing. Learning to sing is an activity that benefits from the involvement of an instructor. A singer does not hear the same sounds inside his or her head that others hear outside. Therefore, having a guide who can tell a student what kinds of sounds he or she is producing guides a singer to understand which of the internal sounds correspond to the desired sounds required by the style of singing the student aims to re-create.

Extending vocal range

An important goal of vocal development is to learn to sing to the natural limits of one's vocal range without any obvious or distracting changes of quality or technique. Vocal pedagogists teach that a singer can only achieve this goal when all of the physical processes involved in singing (such as laryngeal action, breath support, resonance adjustment, and articulatory movement) are effectively working together. Most vocal pedagogists believe in coordinating these processes by (1) establishing good vocal habits in the most comfortable tessitura of the voice, and then (2) slowly expanding the range.

There are three factors that significantly affect the ability to sing higher or lower: the energy factor, the space factor and the depth factor. These three factors can be expressed in three basic rules: As you sing higher, you must use more energy; as you sing lower, you must use less. As you sing higher, you must use more space; as you sing lower, you must use less. As you sing higher, you must use more depth; as you sing lower, you must use less.

Breathing and breath support

Natural breathing has three stages: a breathing-in period, a breathing out period, and a resting or recovery period; these stages are not usually consciously controlled. Within singing, there are four stages of breathing: a breathing-in period (inhalation); a setting up controls period (suspension); a controlled exhalation period (phonation); and a recovery period. These stages must be under conscious control by the singer until they become conditioned reflexes. Many singers abandon conscious controls before their reflexes are fully conditioned which ultimately leads to chronic vocal problems.

Vibrato

Vibrato is a technique in which a sustained note wavers very quickly and consistently between a higher and a lower pitch, giving the note a slight quaver. Vibrato is the pulse or wave in a sustained tone. Vibrato occurs naturally and is the result of proper breath support and a relaxed vocal apparatus. Some singers use vibrato as a means of expression. Many successful artists can sing a deep, rich vibrato.

Extended vocal technique

Extended vocal techniques include rapping, screaming, growling, overtones, falsetto, yodeling[3], using a microphone and sound system, among others.

III. Vocal Music

Vocal music is music performed by one or more singers, which are typically called songs, and which may be performed with or without instrumental accompaniment, in which singing provides the main focus of the piece. Vocal music is probably the oldest form of music since it does not require any instrument or equipment besides the voice. All musical cultures have some form of vocal music and there are many long-standing singing traditions throughout the world's cultures. Music which employs singing but does not feature it prominently is generally considered as instrumental music. Vocal music typically features sung words called lyrics, although there are notable examples of vocal music that are performed using non-linguistic syllables or noises, sometimes as musical onomatopoeia. A short piece of vocal music with lyrics is broadly termed a song, although, in classical music, terms such as aria are typically used.

Genres of vocal music

Vocal music is written in many different forms and styles which are often labeled within a particular genre of music. These genres include Indian classical music, Art music, popular music, traditional music, regional and national music, and fusions of those genres. Within these larger genres are many subgenres. For example, popular music would encompass blues, jazz, country music, easy listening, hip hop, rock music, and several other genres. There may also be a subgenre within a subgenre such as vocalese and scat singing in jazz.

Popular music

The main characteristic of popular singing is that it is easy to sing, so it is easy to be popular. With strong improvisation, popular singing can be sung freely and it emphasizes natural voice, sincere feelings and prominent personality. In many modern pop musical groups, a lead singer performs the primary vocals or melody of a song, as opposed to a backing singer who sings backup vocals or the harmony of a song. Backing vocalists sing some, but usually not all, parts of the song often singing only in a song's refrain or humming in the background. An exception is five-part gospel a cappella music, where the lead is the highest of the five voices and sings a descant and not the melody. Some artists may sing both the lead and backing vocals on audio recordings by overlapping recorded vocal tracks.

Popular music includes a range of vocal styles. Hip-hop uses rapping, the rhythmic delivery of rhymes in a rhythmic speech over a beat or without accompaniment. Some types of rapping consist mostly or entirely of speech and chanting, like the Jamaican "toasting". In some types of rapping, the performers may interpolate short sung or half-sung passages. Blues singing is based on the use of the blue notes—notes sung at a slightly lower pitch than that of the major scale for expressive purposes. In heavy metal and hardcore punk subgenres, vocal styles can include techniques such as screams, shouts, and unusual sounds such as the "death growl". One difference between live performances in the popular and Classical genres is that whereas Classical performers often sing without amplification in small- to mid-size halls, in popular music, a microphone and PA system (amplifier and speakers) are used in almost all performance venues, even a small coffee house.

IV. Bel Canto

Bel canto (Italian for "beautiful singing" or "beautiful song") is a style of opera or opera singing in the 19th century in which producing a beautiful tone was considered very important. The phrase was not associated with a "school" of singing until the middle of the 19th century, when writers in the early 1860s used it nostalgically to describe a manner of singing that had begun to wane around 1830. Nonetheless, neither musical nor general dictionaries attempted to define bel canto until after 1900. The term remains vague and ambiguous in the 21st century and is often used to evoke a lost singing tradition.

History of bel canto and its various definitions

The 18th and early 19th centuries

As generally understood today, the term bel canto refers to the Italian-originated vocal style that prevailed throughout most of Europe during the 18th and early 19th centuries which musicologists sometimes call the "bel canto era". The music of Handel and his contemporaries, as well as that of Mozart and Rossini, benefits from an application of bel canto principles.

The Harvard Dictionary of Music by Willi Apel says that bel canto denotes "the Italian vocal technique of the 18th century, with its emphasis on beauty of sound and brilliance of performance rather than dramatic expression or romantic emotion. In spite of the repeated reactions against bel canto and the frequent exaggeration of its virtuoso element (coloratura), it must be considered as a highly artistic technique and the only proper one for Italian opera and for Mozart. Its early development is closely bound up with that of the Italian opera seria".

19th-century Italy, Germany and France

The popularity of the bel canto style as espoused by Rossini, Donizetti and Bellini faded in Italy during the mid-19th century. It was overtaken by a heavier, more ardent, less embroidered approach to singing that was necessary to perform the innovative works of Giuseppe Verdi (1813–1901) with maximum dramatic impact.

The phrase "bel canto" was not commonly used until the latter part of the 19th century, when it was set in opposition to the development of a weightier, more powerful style of speech-inflected singing associated with German opera and, above all, Richard Wagner's revolutionary music dramas. Wagner (1813–1883) decried the Italian singing model, and he advocated a new, Germanic school of singing that would draw the spiritually energetic and profoundly passionate into the orbit of its matchless Expression.

French musicians and composers never embraced the more florid extremes of the 18th-century Italian bel canto style. They disliked the castrato voice and because they emphasized the clear enunciation of the texts of their vocal music, they objected to the sung word being obscured by excessive fioritura.

Detractors

One reason for the eclipse of the old Italian singing model was the growing influence within the music world of bel canto's detractors, who considered it to be outmoded and condemned it as vocalization lack of content. In the late-19th century and early-20th century, the term "bel canto" was resurrected by singing teachers in Italy.

During the 1890s, the directors of the Bayreuth Festival4 initiated a particularly forceful style of Wagnerian singing that was totally different from the bel canto style. Called "Sprechgesang" by its proponents, the new Wagnerian style prioritized articulation of the individual words of the composer's libretti over legato delivery. This text-based, anti-legato approach to vocalism spread across the German-speaking parts of Europe prior to World War I.

As a result of these many factors, the concept of bel canto became shrouded in mystique and confused by a lot of individual notions and interpretations. Since the singing style of later 17th-century Italy did not differ from that of the 18th century and early 19th century, a connection can be drawn; but, most musicologists agree that the term is best limited to its mid-19th-century use, designating a style of singing that emphasized beauty of tone and technical expertise in the delivery of music that was either highly florid or featured long, flowing and difficult-to-sustain passages of cantilena.

Revival

In the 1950s, the phrase "bel canto revival" was coined to refer to a renewed interest in the operas of Donizetti, Rossini and Bellini. These composers had begun to go out of fashion during the latter years of the 19th century and their works, while never completely disappearing from the performance repertoire, were staged infrequently during the first half of the 20th century, when the operas of Wagner, Verdi and Puccini held sway. That situation changed significantly after World War II with the advent of a group of enterprising orchestral conductors and the emergence of a fresh generation of singers who had acquired bel canto techniques. These artists brought new life into Donizetti, Rossini and Bellini's stage compositions, treating them seriously as music and re-popularizing them throughout Europe and America. Many 18th-century operas that require adroit bel canto skills have also experienced post-war revivals, ranging from lesser-known Mozart and Haydn to extensive Baroque works by Handel, Vivaldi and others.

The main features of the bel canto style:
- Prosodic singing (use of accent and emphasis).
- Matching register and tonal quality of the voice to the emotional content of the words.
- A highly articulated manner of phrasing based on the insertion of grammatical and rhetorical pauses.
- A delivery varied by several types of legato and staccato.
- A liberal application of more than one type of portamento.
- Messa di voce as the principal source of expression.
- Frequent alteration of tempo through rhythmic rubato and the quickening and slowing of the overall time.
- The introduction of a wide variety of graces and divisions into both arias and recitatives.
- Gesture as a powerful tool for enhancing the effect of the vocal delivery.
- Vibrato primarily reserved for heightening the expression of certain words and for gracing longer notes.

Teaching legacy

During the late 18th century and the early 19th century, singing teachers imparted "bel canto technique" to their students. Many famous 18th-century vocal teachers were castrati who contributed a lot to the development of European vocal music and laid the foundation of bel canto. All their pedagogical works follow the same structure, beginning with exercises on single notes and eventually progressing to scales and improvised embellishments. The really creative ornamentation required for cadenzas, involving models and formulae that could generate newly improvised material, came towards the end of the process. Today's pervasive idea that singers should refrain from improvising and always adhere strictly to the letter of a composer's published score is a comparatively recent phenomenon.

Early 19th-century teachers described the voice as being made up of three registers. The chest register was the lowest of the three and the head register the highest, with the passaggio in between. These registers needed to be smoothly blended and fully equalized before a trainee singer could acquire total command of his or her natural instrument, and the surest way to achieve this outcome was for the trainee to practice vocal exercises assiduously. Bel canto—era teachers were great believers in the benefits of vocalise and solfeggio. They strove to strengthen the respiratory muscles of their pupils and equip them with such time-honoured vocal attributes as purity of tone, perfection of legato, phrasing informed by eloquent portamento, and exquisitely turned ornaments.

Major refinements occurred to the existing system of voice classification during the 19th century as the international operatic repertoire diversified, split into distinctive nationalist schools and expanded in size. Whole new categories of singers such as mezzo-soprano and Wagnerian bass-baritone arose towards the end of the 19th century, as did such new sub-categories as lyric coloratura soprano, dramatic soprano and spinto soprano, and various grades of tenor, stretching from lyric through spinto to heroic. These classificatory changes have had a lasting effect on how singing teachers designate voices and opera house5 managements cast productions.

There was, however, no across-the-board uniformity among 19th-century bel canto adherents in passing on their knowledge and instructing students. Each had their own training regimes and pet notions. Fundamentally, though, they all subscribed to the same set of bel canto precepts, and the exercises that they devised to enhance breath support, dexterity, range, and technical control remain valuable and, indeed, some teachers still use them.

V. Musical Classics

Burning Flame	《燃烧的火焰》
Na Buguo	《那布果》
Requiem	《安魂曲》
Tosca	《托斯卡》
Country Knight	《乡村骑士》

La Traviata	《茶花女》
Onegin	《奥涅金》
Carmen	《卡门》
Madame Butterfly	《蝴蝶夫人》
Roaring Up to the Sky	《仰天长啸》
Thunderstorm	《雷雨》
On the Songhua River	《松花江上》
ManJianghong	《满江红》
Star of Life	《生命的星》
River	《河流》
Romance of the Western Chamber	《西厢记》
You Are Such a Person	《你是这样的人》
I Come from the Sky like Snowflakes	《我像雪花天上来》
Kashgar Girl	《喀什葛尔女郎》
See Siberia Again	《又见西柏坡》

Notes on the text

1. Vocal break 换声
2. Vocal weight 声音重量
3. yodeling 用约德尔唱法歌唱；用真假嗓音交替歌唱
4. the Bayreuth Festival 拜罗伊特音乐节。拜罗伊特音乐节是德国的一项音乐节日，第一届拜罗伊特音乐节于1876年举办。
5. opera house 歌剧院

Terms

larynx	/ˈlærɪŋks/	喉；咽喉
vocal cord (vocal fold)	/ˈvəʊkl kɔːrd/	声带
epiglottis	/ˌepɪˈglɑːtɪs/	会厌；喉头盖
vocal tract	/ˈvəʊkl trækt/	声道
nasal passage	/ˈneɪzl ˈpæsɪdʒ/	鼻道；鼻腔通道
soft palate	/sɔːft ˈpælət/	软腭
register	/ˈredʒɪstər/	声区
vocal fold mucosa	/ˈvəʊkl fəʊld mjuˈkoʊsə/	声襞粘膜
respiration	/ˌrespəˈreɪʃn/	呼吸
phonation	/foʊˈneɪʃən/	发声

inhalation	/ˌɪnhəˈleɪʃn/	吸入
suspension	/səˈspenʃn/	控制气息
resonation	/ˈrezəneɪʃn/	共鸣
articulation	/ɑːrˌtɪkjuˈleɪʃn/	咬字吐字；清楚的咬字（或发音）
the vocal fry register	/ðiː ˈvəʊkl fraɪ ˈredʒɪstər/	气泡音 M0
the modal register/chest voice	/ðiː ˈməʊdl ˈredʒɪstər/, /tʃest vɔɪs/	胸声 M1
the falsetto register/ head voice	/ðiː fɔːlˈsetəʊ ˈredʒɪstər/, /hed vɔɪs/	头声 M2
the whistle register	/ðiː ˈwɪsl ˈredʒɪstər/	海豚音 M3
aria	/ˈɑːriə/	咏叹调
recitative	/ˌresɪtəˈtiːv/	宣叙调
da capo arias	/ˌdɑːˈkɑːpəʊ ˈɑːriəz/	返始咏叹调；再现咏叹调
arranged	/əˈreɪndʒd/	改编的；谱曲的
improvised	/ˈɪmprəvaɪzd/	即兴的
lyric	/ˈlɪrɪk/	歌词
audition	/ɔːˈdɪʃn/	试镜；海选
voice teacher (or vocal coach)	/vɔɪs ˈtiːtʃər/	声乐导师
lead singer	/liːd ˈsɪŋər/	主唱；领唱
backing singer	/ˈbækɪŋ ˈsɪŋər/	伴唱；和声歌手
castrati	/kæˈstrɑːtiː/	阉人歌手（castrato 的复数）
spinto soprano	/ˈspɪntoʊ səˈprɑːnəʊ/	抒情兼戏剧女高音
coloratura	/ˌkʌlərəˈtʊrə/	花腔；花腔女高音歌手
humming	/ˈhʌmɪŋ/	哼唱
lip-syncing	/ˈlɪp sɪŋkɪŋ/	对口型；假唱
vocalize	/ˈvəʊkəlaɪz/	发声；练声
vibrato	/vɪˈbrɑːtəʊ/	颤音
rapping	/ˈræpɪŋ/	饶舌
overtone	/ˌəʊvərtəʊn/	泛音
falsetto	/fɔːlˈsetəʊ/	假声
yodeling	/ˈjoʊdlɪŋ/	约德尔唱法；用真假嗓音交替歌唱
solfeggio	/sɒlˈfedʒoʊ/	视唱练耳
cadenza	/kəˈdenzə/	华彩乐段；装饰乐段

ornamentation	/ˌɔːrnəmenˈteɪʃn/	装饰
ornament	/ˈɔːrnəmənt/	装饰音
fioritura	/fiˌɔːrɪˈtʊrə/	装饰音；花音
portamento	/ˌpoʊrtəˈmentoʊ/	延音；滑音
Sprechgesang	/ˈʃprexɡəzæŋ/	朗诵唱；道白式演唱法
cappella	/kəˈpelə/	无伴奏（合唱）

Exercises

I. Comprehension questions

1. What is the definition of singing?
2. Who usually provides voice training for professional singers throughout their careers?
3. What is the definition of vocal registration?
4. How many different major voice categories are there within classical music?
5. What is Wagner's view to bel canto style? And what does he advocate?

II. Translating useful expressions

1. 演唱可以是正式的或非正式的，谱曲的或即兴的。
2. 在很多现代流行音乐中，主唱负责歌曲的旋律部分，伴唱则负责歌曲的和声部分。
3. 美声唱法被认为是最科学的、极具艺术技巧的唱法。
4. The sound of each individual's singing voice is entirely unique not only because of the actual shape and size of an individual's vocal cords, but also due to the size and shape of the rest of that person's body.
5. Vocal coaches are great believers in the benefits of vocalise and solfeggio.
6. Once the auditions conclude, coaches have their team of artists and the competition begins. Coaches then mentor their artists and they compete to find the best singer.

III. Brainstorm

1. Vocal music is probably the oldest form of music since it does not require any instrument or equipment besides the voice. All musical cultures have some form of vocal music and there are many long-standing singing traditions

throughout the world's cultures. Please research Chinese national singing style, and then brainstorm its difference with bel canto.

2. There are several well-known television shows that showcase singing, such as *Super Girl, The Voice etc*. Please research the competition rules of these games, and present them in class. Brainstorm the following questions: Would you like to be famous? If you make it, how would you give back to the society?

Appendix II   Glossary

I. Chinese Musical Instruments

acciaccatura	/ɑˌtʃakəˈturə/	打音
acrylic/artificial nail	/əˈkrɪlɪk neɪl/, /ˌɑːrtɪˈfɪʃl neɪl/	假指甲
air hole	/er hoʊl/	按孔
alto	/ˈæltoʊ/	中音；女低音
alto bridge	/ˈæltoʊ brɪdʒ/	中音马条
amplification pipe	/ˌæmplɪfɪˈkeɪʃn paɪp/	共鸣管
arpeggio	/ɑːrˈpedʒioʊ/	琶音
artificial harmonics	/ˌɑːrtɪˈfɪʃl hɑːrˈmɑːnɪks/	人工泛音
back nut	/bæk nʌt/	后岳山
base	/beɪs/	笙斗
baseboard	/ˈbeɪsbɔːrd/	底板
bass	/beɪs/	低音
bass bridge	/beɪs brɪdʒ/	低音马条
bell	/bel/	喇叭
blowing hole	/bloʊɪŋ hoʊl/	吹孔
board	/bɔːrd/	面板
bocal	/bokal/	芯子
bridge	/brɪdʒ/	琴马（筌篌）；雁柱（古筝）；岳山（古琴）
bridge pin	/brɪdʒ pɪn/	弦钉
brushing	/ˈbrʌʃɪŋ/	扫弦
ceremonial cap	/ˌserɪˈmoʊniəl kæp/	冠角；焦尾
cheng lu		承露
chordal appoggiature	/ˈkɔːdəl apɔ(d)ʒjatyːr/	抹音
chromatic bridge	/krəˈmætɪk brɪdʒ/	半音马条
circle breathing	/ˈsɜːrkl ˈbriːðɪŋ/	循环换气
circular breathing/circle breathing	/ˈsɜːrkjələr ˈbriːðɪŋ/, /ˈsɜːrkl ˈbriːðɪŋ/	循环换气法

column/pillar	/ˈkɑːləm/, /ˈpɪlər/	立柱
cylindrical nut	/səˈlɪndrɪkl nʌt/	变音槽
détaché	/detaʃe/	分弓
double notes	/ˈdʌbl noʊts/	双竹
double strings	/ˈdʌbl strɪŋz/	双音；双弹
double tonguing	/ˈdʌbl ˈtʌŋɪŋ/	双吐
down & up stroke	/daʊn ən ʌp stroʊk/	弹挑
down portamento	/daʊn ˌpoʊrtəˈmentoʊ/	下滑音
down-sliding	/daʊn ˈslaɪdɪŋ/	注
down-stroke brushing	/daʊn stroʊk ˈbrʌʃɪŋ/	扫
dragon pool	/ˈdræɡən puːl/	龙池
dragon's gum/nut	/ˈdræɡənz ɡʌm/, /ˈdræɡənz nʌt/	龙龈
ensemble	/ɑːnˈsɑːmbl/	重奏
face plate	/feɪs pleɪt/	面板
false nail	/fɔːls neɪl/	假指甲
fast bow	/fæst baʊ/	快弓
fine tuner	/faɪn ˈtuːnər/	滚轴
finger hole	/ˈfɪŋɡər hoʊl/	指孔
finger pick	/ˈfɪŋɡər pɪk/	义甲
fingerboard	/ˈfɪŋɡərbɔːrd/	指板
flowing water/ repeating glissando	/ˈfloʊɪŋ ˈwɔːtər/, /rɪˈpiːtɪŋ ɡlɪˈsændoʊ/	流水
flutter tonguing	/ˈflʌtər ˈtʌŋɪŋ/	花舌
flying finger trill	/ˈflaɪɪŋ ˈfɪŋɡər trɪl/	飞指
foot	/fʊt/	琴足
free rhythm	/friː ˈrɪðəm/	散板
fret	/fret/	品
front nut	/frʌnt nʌt/	前岳山
glide/ portamento	/ɡlaɪd/, /ˌpoʊrtəˈmentoʊ/	滑音
glissando	/ɡlɪˈsændoʊ/	刮奏；历音
gum supporter	/ɡʌm səˈpɔːrtər/	龈托
hammer	/ˈhæmər/	琴竹
handle of modulation	/ˈhændl əv ˌmɑːdʒəˈleɪʃn/	转调手柄
harmonics/ overtone	/hɑrˈmɑnɪks/, /ˈoʊvərtoʊn/	泛音
head	/hed/	凤头
head/ top/ tip of neck	/hed əv nek/, /tɑːp əv nek/, /tɪp əv nek/	琴头
hole for penetrating string	/hoʊl fɔːr ˈpenətreɪtɪŋ strɪŋ/	穿弦孔

index finger	/ˈɪndeks ˈfɪŋɡər/	食指
inner string	/ˈɪnər strɪŋ/	内弦
legato	/lɪˈɡɑːtəʊ/	连弓
lower acciaccature	/ˈləʊər ɑːˌtʃɑːkɑːˈtʊərə/	打音
lute head	/luːt hed/	琴头
marker	/ˈmɑːrkər/	徽位
membrane	/ˈmembreɪn/	笛膜
middle finger	/ˈmɪdl ˈfɪŋɡər/	中指
mount yue / bridge	/brɪdʒ/	岳山
mouthpiece	/ˈmaʊθpiːs/	笙嘴
muffling	/ˈmʌflɪŋ/	止音
mute	/mjuːt/	伏／煞音
natural harmonics	/ˈnætʃrəl hɑːrˈmɑːnɪks/	自然泛音
neck	/nek/	凤项；琴颈
neck/handle	/nek/, /ˈhændl/	琴杆
non-pitch tone	/ˈnoʊn pɪtʃ toʊn/	弦外音
note end acciaccature	/nəʊt end ɑːˌtʃɑːkɑːˈtʊərə/	赠音
note window	/nəʊt ˈwɪndəʊ/	音窗
nut	/nʌt/	山口（西方乐器：琴枕）
nut/ looping cord	/nʌt kɔːrd/, /ˈluːpɪŋ kɔːrd/	千斤
open tone	/ˈəʊpən təʊn/	散音
outer string	/ˈaʊtər strɪŋ/	外弦
panel/sound board	/ˈpænl/, /saʊnd bɔːrd/	面板；共鸣板
panel/ board	/ˈpænl/, /bɔːrd/	面板
peg hole/ peg box	/peg hoʊl/, /peg bɑːks/	弦槽
peg shield/ protector	/peg ʃiːld/, /peg prəˈtektər/	护枕
pegbox	/ˈpegbɔks/	弦轴箱
pentatonic scale	/ˌpentəˈtɑːnɪk skeɪl/	五声音阶
phoenix forehead	/ˈfiːnɪks ˈfɔːrhəd/	凤额
phoenix pool	/ˈfiːnɪks puːl/	凤池
piccolo	/ˈpɪkələʊ/	短笛
pinch	/pɪntʃ/	掐音
pinky finger	/ˈpɪŋki ˈfɪŋɡər/	小指
pipe	/paɪp/	笙苗
pirouette	/ˌpɪruˈet/	气盘
pizzicato	/ˌpɪtsɪˈkɑːtəʊ/	拨弦
plectrum/ single pick	/ˈplektrəm/, /ˈsɪŋɡl pɪk/	琴拨；弹片

pluck	/plʌk/	弹拨
polyphonic	/ˌpɑːliˈfɑːnɪk/	复调
popped note	/pɑːpt nəʊt/	剁音
portamento	/ˌpoʊrtəˈmentoʊ/	滑音
position	/pəˈzɪʃn/	把位
posterior mountain	/pɑːˈstɪriər ˈmaʊntn/	后岳山
press	/pres/	按音
protrusions	/prəˈtruːʒənz/	纳音
pull-push vibrato	/pʊl pʊʃ vɪˈbrɑːtəʊ/	吟弦
pushing and pulling	/ˈpʊʃɪŋ ən pʊlɪŋ/	推拉
python skin	/ˈpaɪθɑːn skɪn/	琴皮；蛇皮
rapid tonguing	/ˈræpɪd ˈtʌŋɪŋ/	碎吐
red wax	/red wæks/	红蜡
reed	/riːd/	哨子；簧片；笙簧
ring finger	/rɪŋ ˈfɪŋɡər/	无名指
scale	/skeɪl/	音阶
scale with harmony/ parallel notes	/skeɪl wɪθ ˈhɑːrməni/, /ˈpærəlel noʊts/	和声音阶
scroll	/skrəʊl/	琴头
sforzandissimo flutter tonguing	/sfɔːtˈsændoʊ ˈflʌtər ˈtʌŋɪŋ/	爆花舌
S-shaped panel/ board	/es ʃeɪpt ˈpænl/, /es ʃeɪpt bɔːrd/	S 形面板
short tremolo	/ʃɔːrt ˈtreməloʊ/	滚奏
shoulder	/ˈʃəʊldər/	仙人肩
side board	/saɪd bɔːrd/	侧板
single note	/ˈsɪŋɡl nəʊt/	单竹
single tonguing	/ˈsɪŋɡl ˈtʌŋɪŋ/	单吐
slap	/slæp/	拍
sliding	/ˈslaɪdɪŋ/	滑音
soft double tonguing	/sɔːft ˈdʌbl ˈtʌŋɪŋ/	软双吐
soprano	/səˈprænəʊ/	高音；女高音
sound hole	/saʊnd həʊl/	出音孔
sound-box	/saʊnd bɑːks/	共鸣箱
sound box/resonator body	/saʊnd bɑːks/, /ˈrezəneɪtər ˈbɑːdi/	琴筒
spiccatto	/spɪˈkɑːtəʊ/	跳弓
staccato	/stəˈkɑːtəʊ/	顿音
stopped tone	/stɑːpt təʊn/	按音
strike	/straɪk/	弹
strike and pluck	/straɪk ən plʌk/	打带音

striking	/ˈstraɪkɪŋ/	打
string	/strɪŋ/	琴弦
string hole	/strɪŋ həʊl/	弦眼
strong vibrato	/strɔːŋ vɪˈbrɑːtəʊ/	气震音
sul ponticello	/ˈsʌl pɒntɪˈtʃeloʊ/	接近琴马演奏
sweep	/swiːp/	扫弦
tailpiece	/ˈteɪlpiːs/	缚弦；系弦板
tangle strings	/ˈtæŋgl strɪŋz/	绞弦
tenor	/ˈtenər/	次中音；男高音
tenor bridge	/ˈtenər brɪdʒ/	次中音条马
thumb	/θʌm/	大拇指
thumb position	/θʌm pəˈzɪʃn/	大指把位
tonguing	/tʌŋ ɪŋ/	吐音
transverse	/ˈtrænzvɜːrs/	横向的
treble bridge	/ˈtenər brɪdʒ/	高音条马
tremolo	/ˈtremələʊ/	抖弓；轮指
tremolo by tonguing	/ˈtremələʊ baɪ tʌŋɪŋ/	呼舌
trill	/trɪl/	（指）颤音
triple tonguing	/ˈtrɪpl ˈtʌŋɪŋ/	三吐
tube	/tuːb/	哨管
tuning box	/ˈtuːnɪŋ bɑːks/	调音盒
tuning peg	/ˈtuːnɪŋ peg/	琴轴
tuning pin	/ˈtuːnɪŋ pɪn/	弦钉（古筝）
tuning screw	/ˈtuːnɪŋ skruː/	弦钉（扬琴）
turn	/tɜːrn/	回滑音
up portamento	/ʌp ˌpoʊrtəˈmentoʊ/	上滑音
up-back sliding	/ʌp bæk ˈslaɪdɪŋ/	进复
upper acciaccature	/ˈʌpər ɑːˌtʃɑːkɑːˈtʊərə/	叠音
up-sliding	/ʌp ˈslaɪdɪŋ/	绰
up-stroke brushing	/ʌp strəʊk ˈbrʌʃɪŋ/	拂
vertical	/ˈvɜːrtɪkl/	竖向的
vibrato	/vɪˈbrɑːtəʊ/	颤音；腹震音；揉弦
waist	/weɪst/	龙腰

II. Western Musical Instruments

accel (abbr. of accelerado)	/əkˈsɛl/	加快的
alberti bass/ broken chord	/alˈbɛrti beɪs/ , /ˈbrəʊkən kɔːrd/	分解和弦
accent	/ækˈsent/	重音
accidental key (black key)	/ˌæksɪˈdentl kiː/	半音键
action	/ˈækʃn/	击弦机
adagio	/əˈdɑːdʒiəʊ/	从容的；慢板
additive synthesis	/ˈædətɪv ˈsɪnθəsɪs/	加法合成
aerophone	/ˈeroʊˌfoʊn/	气鸣乐器
aftertouch	/ˈæftər tʌtʃ/	触后
algorithm	/ˈælgərɪðəm/	算法
allegro	/əˈlegrəʊ/	快板
alternating bass	/ˈɔːltərneɪtɪŋ beɪs/	交替低音
alto clef	/ˈæltəʊ klef/	中音谱号
amplitude	/ˈæmplɪtuːd/	幅度
analog synthesis	/ˈænəlɔːg ˈsɪnθəsɪs/	模拟合成器
andante	/ænˈdænteɪ/	行板
aria	/ˈɑːriə/	咏叹调
arm movement	/ɑːrm ˈmuːvmənt/	手臂动作
arm-swing exercise	/ɑːrm swɪŋ ˈeksərsaɪz/	吊臂练习
arpeggiator	/ɑːrˈpedʒieɪtə/	琶音器
arpeggio	/ɑːrˈpedʒiəʊ/	琶音
arranged	/əˈreɪndʒd/	改编的；谱曲的
articulation	/ɑːrˌtɪkjuˈleɪʃn/	咬字吐字；清楚的咬字（或发音）
attack	/əˈtæk/	起音
attenuator	/əˈtenjueɪtər/	消音器；衰减器
audition	/ɔːˈdɪʃn/	试镜；海选
backing singer	/ˈbækɪŋ ˈsɪŋər/	伴唱；和声歌手
band-pass filter	/bænd pæs ˈfɪltər/	带通滤波器/滤波模式
barrel	/ˈbærəl/	调节管；二节管；脖管
base	/beɪs/	底盘
bass	/beɪs/	贝司
bass bar	/beɪs bɑːr/	音梁；低音梁
bass button	/beɪs ˈbʌtn/	贝司键钮
bass clef	/beɪs klef/	低音谱号
bass drum	/beɪs drʌm/	大鼓

bass joint/ long joint	/beɪs dʒɔɪnt/, /lɔːŋ dʒɔɪnt/	长管
bassist	/ˈbeɪsɪst/	低音提琴手
bassoonist	/bəˈsuːnɪst/	大管演奏者/巴松管手
batter head/ top head	/ˈbætər hed/, /tɑːp hed/	鼓皮/打击面
bell	/bel/	喇叭口
bell joint	/bel dʒɔɪnt/	上节管
bellows shake	/ˈbeloʊz ʃeɪk/	风箱颤音
bellows	/ˈbeloʊz/	风箱
big leap	/bɪg liːp/	大跳
body	/ˈbɑːdi/	琴身
body side	/ˈbɑːdi saɪd/	侧板
bongo	/ˈbɑːŋgoʊ/	邦戈鼓
bore	/bɔːr/	内径；孔径
bottom deck	/ˈbɑːtəm dek/	背板
bow	/baʊ/	弯管（U形管）
bow grip	/baʊ grɪp/	握弓
bowed stringed instrument	/boʊd ˈstrɪŋd ˈɪnstrəmənt/	弓弦乐器
bowing	/ˈboʊɪŋ/	弓法
braccio	/brɑːtʃoʊ/	（意）手臂
brace	/breɪs/	拉管支杆
bridge	/brɪdʒ/	琴马
brush	/ˈbrʌʃ/	鼓刷
bulging	/ˈbʌldʒɪŋ/	鼓起的
Burton grip	/ˈbɜːrtn grɪp/	巴顿握槌法
cadenza	/kəˈdenzə/	华彩乐段；装饰乐段
cappella	/kəˈpelə/	无伴奏（合唱）
carrier wave	/ˈkæriər weɪv/	载波
cast iron frame (plate)	/ˈkæst ˈaɪərn freɪm/	铸铁支架
caster	/ˈkæstər/	脚轮
castrati	/kæˈstrɑːtiː/	阉人歌手（castrato的复数）
chamber music	/ˈtʃeɪmbər ˈmjuːzɪk/	室内乐
chin rest	/tʃɪn rest/	腮托
chord	/kɔːrd/	和弦
chromatic	/krəˈmætɪk/	半音的
circular breathing	/ˈsɜːrkjələr ˈbriːðɪŋ/	循环呼吸
clarinet	/ˌklærəˈnet/	单簧管（又称竖笛、黑管）
clarinetist	/ˌklærəˈnetɪst/	单簧管演奏者
closed roll	/kloʊzd roʊl/	密集型滚奏

col legno		敲弓
coloratura	/ˌkʌləˈtʊrə/	花腔；花腔女高音歌手
compass	/ˈkʌmpəs/	音域
concerto	/kənˈtʃɜːrtoʊ/	协奏曲
concerto grosso	/kənˈtʃɜːrtoʊ ˈgroʊsoʊ/	大协奏曲
conga	/ˈkɑːŋgə/	康加鼓
conical tube	/ˈkɑːnɪkl tuːb/	圆锥管
console	/kənˈsoʊl/	控制台
contrabass	/ˈkɑːntrəbeɪs/	（倍）低音的
contrabassoon/double bassoon	/ˈkɑːntrəbəsuːn/, /ˈdʌbl bəˈsuːn/	低音大管
control voltage	/kənˈtroʊl ˈvəʊltɪdʒ/	控制电压
cork	/kɔːrk/	软木塞
counterweight	/ˈkaʊntərweɪt/	平衡器
credential	/krəˈdenʃl/	文凭；资格
crescendo	/krəˈʃendoʊ/	渐强
crook	/krʊk/	变音插管；定调管；S管
crossing hand	/ˈkrɔːsɪŋ hænd/	双手交叉
cross-stick	/ˈkrɔːs stɪk/	止音边击
crown	/kraʊn/	笛头塞；软木塞子
cutoff frequency	/ˈkʌt ɔːf ˈfriːkwənsi/	截止频率
cutoff point	/ˈkʌt ɔːf pɔɪnt/	截止点
cylindrical	/səˈlɪndrɪkl/	圆柱形的
da capo arias	/ˌdɑː ˈkɑːpoʊ ˈɑːriəz/	返始咏叹调；再现咏叹调
damper	/ˈdæmpər/	制音器
dead stroke	/ded stroʊk/	压奏
debate	/dɪˈbeɪt/	争论
decay	/dɪˈkeɪ/	衰减
decrescendo	/ˌdiːkrəˈʃendoʊ/	减弱
delay	/dɪˈleɪ/	混响延时/延迟
descendant	/dɪˈsendənt/	后代；派生物
detached	/dɪˈtætʃt/	分弓
diaphragmatic breathing	/daɪəfrægˈmætɪk ˈbriːðɪŋ/	腹式呼吸
diatonic	/ˌdaɪəˈtɑːnɪk/	自然音的
double action	/ˈdʌbl ˈækʃn/	双重作用
double joint/ boot joint/ butt joint	/ˈdʌbl dʒɔɪnt/, /buːt dʒɔɪnt/, /bʌt dʒɔɪnt/	底管
double reed	/ˈdʌbl riːd/	双簧
double stroke	/ˈdʌbl stroʊk/	双跳

double tonguing	/ˈdʌbl ˈtʌŋɪŋ/	双吐
downstroke	/ˈdaʊnˌstroʊk/	弹
duple/ triple/ quadruple bellows shakes	/ˈduːpəl ˈbeloʊz ʃeɪks/, /ˈtrɪpl ˈbeloʊz ʃeɪks/, /kwɑːˈdruːpl ˈbeloʊz ʃeɪks/	双重 / 三重 / 四重风箱颤音
duration	/duˈreɪʃn/	时值
embouchure	/ˌɑːmbʊˈʃʊr/	吹口；口型
embouchure hole	/ˌɑːmbʊˈʃʊr həʊl/	吹孔
employ	/ɪmˈplɔɪ/	使用
endpin	/ˈendpɪn/	琴脚；尾柱
envelope	/ˈɑːnvələʊp/	包络
envelope genetator	/ˈɑːnvələʊp ˈdʒenəreɪtər/	包络发生器
epiglottis	/ˌepɪˈɡlɑːtɪs/	会厌；喉头盖
expression pedal	/ɪkˈspreʃn ˈpedl/	表情踏板
F attachment	/ef əˈtætʃmənt/	(转调键控制的) 滑管
F hole	/ef həʊl/	孔；音孔
F lever	/ef ˈlevər/	变音键；转调键
falsetto	/fɔːlˈsetəʊ/	假声
ferrule/ bumper	/ˈferəl ˈbʌmpər/	橡胶垫
filter	/ˈfɪltər/	滤波器
finger ring	/ˈfɪŋɡər rɪŋ/	指环
fingerboard	/ˈfɪŋɡərbɔːrd/	指板
fingering	/ˈfɪŋɡərɪŋ/	指法
fioritura	/fiˌɔːrɪˈtʊrə/	装饰音；花音
floral tonguing	/ˈflɔːrəl ˈtʌŋɪŋ/	花舌音
flutter tonguing	/ˈflʌtər ˈtʌŋɪŋ/	颤舌音；花舌音
foot switch	/fʊt swɪtʃ/	脚控开关
forearm rotation	/ˈfɔːrɑːrm rəʊˈteɪʃn/	小臂旋转
form	/fɔːrm/	形制
fourth lever	/fɔːrθ ˈlevər/	转调键
frame	/freɪm/	鼓架
free bass	/friː beɪs/	自由低音
frequency modulation (FM) synthesis	/ˈfriːkwənsi ˌmɑːdʒəˈleɪʃn ˈsɪnθəsɪs/	调频（FM）合成器
fret	/fret/	品
fretboard/fingerboard	/ˈfretbɔːrd/, /ˈfɪŋɡərbɔːrd/	指板
fundamental	/ˌfʌndəˈmentl/	基音
glissando	/ɡlɪˈsændəʊ/	刮奏（钢琴）

glissando/ sliding	/glɪˈsændoʊ /, /ˈslaɪdɪŋ/	滑音
glockenspiel	/ˈglɑːkənʃpiːl/	钟琴
gong	/gɔːŋ/	大锣
grave	/greɪv/	慢速
growling	/ˈgraʊlɪŋ/	咆哮音；爆破音
grille	/grɪl/	琴盖；护栅
hammer	/ˈhæmər/	琴槌
hand motion	/hænd ˈmoʊʃn/	手的运动
hand rest/crutch	/hænd rest/, /krʌtʃ/	手托
harmonics	/hɑːrˈmɑːnɪks/	泛音
head/skin	/hed/, /skɪn/	鼓皮
heel	/hiːl/	琴肩
heyday	/ˈheɪdeɪ/	全盛时期
high-pass filter	/haɪ pæs ˈfɪltər/	高通滤波器 / 滤波模式
hi-hat	/ˈhaɪ hæt/	踩镲
hoop/metal hoop	/huːp/, /ˈmetl huːp/	鼓圈
hum at the same time	/hʌm ət ðə seɪm taɪm/	同时哼唱
humming	/ˈhʌmɪŋ/	哼唱
level/ amp/ volume	/ˈlevl/, /æmp/, /ˈvɑːljəm/	音量（控制单独振荡器的音量）
improvised	/ˈɪmprəvaɪzd/	即兴的
inhalation	/ˌɪnhəˈleɪʃn/	吸入
keep fingers on the key		手指贴键
kettle	/ˈketl/	鼓身；鼓桶
key guard	/kiː gɑːrd/	音键保护框架；保护网
keyboard	/ˈkiːbɔːrd/	键盘
keywork	/kiː wɜːrk/	按键
largo	/ˈlɑːrgəʊ/	庄严的慢板
larynx	/ˈlærɪŋks/	喉；咽喉
lead pipe	/liːd paɪp/	号嘴导管
lead singer	/liːd ˈsɪŋər/	主唱；领唱
leadpipe	/ˈliːdpaɪp/	导管
legato	/lɪˈgɑːtəʊ/	连音；连奏；圆滑连弓
lento	/ˈlentəʊ/	广板；广阔地
liberal arts	/ˈlɪbərəl ˈɑːrts/	人文学科
lid	/lɪd/	翻盖
ligature	/ˈlɪgətʃər/	吹口束圈
lip plate (or embouchure plate)	/lɪp pleɪt/	嘴唇贴盘

lip-syncing	/ˈlɪp sɪŋkɪŋ/		对口型；假唱
little finger key	/ˈlɪtl ˈfɪŋgər kiː/		小指键
lower keyboard	/ˈləʊər ˈkiːbɔːrd/		下键盘
low-frequency oscillator (LFO)	/loʊ ˈfriːkwənsi ˈɑːsɪleɪtər/		低频滤波震荡器
low-pass filter	/loʊ pæs ˈfɪltər/		低通滤波器/滤波模式
luthier	/ˈlutiə/		琴师
lyric	/ˈlɪrɪk/		歌词
main slide	/meɪn slaɪd/		主管
marching snare	/ˈmɑːrtʃɪŋ sner/		行进军鼓
matched grip	/mætʃt grɪp/		扣腕握槌法
memory button	/ˈmeməri ˈbʌtn/		注册记忆按钮
middle pedal (a sostenuto/ double weak pedal)	/ˈmɪdl ˈpedl/		延音/消音/倍弱音踏板
MIDI: Musical Instrument Digital Interface	/ˈmɪdi/: /ˈmjuːzɪkl ˈɪnstrəmənt ˈdɪdʒɪtl ˈɪntərfeɪs/		电子音乐数字接口
modulation	/ˌmɑːdʒəˈleɪʃn/		调制；调节；改变
mouthpiece	/ˈmaʊθpiːs/		吹嘴；号嘴；吹口；笛头
multiphonics	/mʌlti ˈfɑːnɪks/		复合音
musette	/mjuːˈzet/		缪塞特音
Musser grip	/myse grɪp/		穆瑟握槌法
mute (or sordino)	/mjuːt/		弱音器
nasal passage	/ˈneɪzl ˈpæsɪdʒ/		鼻道；鼻腔通道
natural key (white key)	/ˈnætʃrəl kiː/		全音键
neck/ crook	/nek/, /krʊk/		弯脖；吹管；颈管；共鸣管；琴颈
neck screw	/nek skruː/		吹管固定螺丝
neckpipe	/ˈnekpaɪp/		颈管
norm	/nɔːrm/		常态
notation	/nəʊˈteɪʃn/		记谱
nut	/nʌt/		琴枕；弦枕
oboist	/ˈəʊbəʊɪst/		双簧管吹奏者
octave	/ˈɑːktɪv/		八度音阶
octave key	/ˈɑːktɪv kiː/		八度音键；高音键
octave pin	/ˈɑːktɪv pɪn/		泛音键联动杆
off the key	/ɔːf ði kiː/		手指离键
open roll	/ˈoʊpən roʊl/		开放型滚奏
ornament	/ˈɔːrnəmənt/		装饰音
ornamentation	/ˌɔːrnəmenˈteɪʃn/		装饰

oscillator (OSC)	/ˈɑːsɪleɪtər/	振荡器
outer slide tube	/ˈaʊtər slaɪd tuːb/	外拉管
overblow	/ˌoʊvəˈbloʊ/	超吹
overblowing	/ˌoʊvərˈbloʊɪŋ/	吹出泛音
overblown note	/ˌoʊvərˈbloʊn noʊt/	吹出泛音
overtone	/ˈəʊvərtəʊn/	泛音
pair cymbals	/per ˈsɪmbəlz/	对镲
paradiddle	/ˌpærəˈdɪdəl/	复合跳
parameter	/pəˈræmɪtər/	参数
pedal	/ˈpedl/	踏板
pedal keyboard	/ˈpedl ˈkiːbɔːrd/	脚键盘
peghead	/ˈpeghed/	琴头
perfect fifth	/ˈpɜːrfɪkt fɪfθ/	纯五度
phonation	/foʊˈneɪʃən/	发声
pitch	/pɪtʃ/	音高调节
pizzicato	/ˌpɪtsɪˈkɑːtəʊ/	拨弦
pluck	/plʌk/	弹拨
plug-in	/ˈplʌɡ ɪn/	插件
portamento	/ˌpoʊrtəˈmentoʊ/	延音；滑音
position	/pəˈzɪʃn/	把位
preamp	/ˈpriæmp/	前置放大器
prestissimo	/presˈtɪsəˌmo/	急板
protector cap (contains the U-tube)	/prəˈtektər kæp/	金属帽（含 U 形管）
pulse wave	/pʌls weɪv/	脉冲波
quarter/ half/ whole note	/ˈkwɔːrtər noʊt/, /hæf noʊt/, /hoʊl noʊt/	四分 / 二分 / 全音符
quartet	/kwɔːrˈtet/	四重奏
quintet	/kwɪnˈtet/	五重奏
rapping	/ˈræpɪŋ/	饶舌
recital	/rɪˈsaɪtl/	独奏 / 唱音乐会
recitative	/ˌresɪtəˈtiːv/	宣叙调
reed	/riːd/	簧片
register	/ˈredʒɪstər/	声区
register key	/ˈredʒɪstər kiː/	泛音键
release	/rɪˈliːs/	释放；尾音
repertoire	/ˈrepərtwɑːr/	保留 / 全部曲目
resonation	/ˈrezəneɪʃn/	共鸣

respiration	/ˌrespəˈreɪʃn/	呼吸
reverberation module	/rɪˌvɜːrbəˈreɪʃn ˈmɑːdʒuːl/	混响模块
rhythm module	/ˈrɪðəm ˈmɑːdʒuːl/	节奏模块
ride cymbal	/raɪd ˈsɪmbəl/	叮叮镲
rim	/rɪm/	（鼓）压圈
rim (case)	/rɪm/	琴箱
rimshot	/ˈrɪm ʃɑːt/	鼓边平击；重音边击
rod system	/rɑːd ˈsɪstəm/	连杆系统
rosette	/rəʊˈzet/	音孔环饰
rotary valve	/ˈrəʊtəri vælv/	转式活塞
rubato	/ruˈbɑtoʊ/	自由速度
run	/rʌn/	急奏
saddle	/ˈsædl/	下弦枕
sample	/ˈsæmpl/	采样；样本
sample-based synthesis	/ˈsæmpl beɪst ˈsɪnθəsɪs/	采样合成
sawtooth wave	/ˈsɔːˌtuːθ weɪv/	锯齿波
scale	/skeɪl/	音阶
scordatura	/ˌskɔːdɑːˈtuːrɑː/	（意）特殊调弦
screw	/skruː/	琴钉
scroll	/skrəʊl/	琴头
sequence	/ˈsiːkwəns/	音序器
sextet	/seksˈtet/	六重奏
shell	/ʃel/	鼓腔
sine wave	/saɪn weɪv/	正玄波
single stroke	/ˈsɪŋgl strəʊk/	单击
single tonguing	/ˈsɪŋgl ˈtʌŋɪŋ/	单吐
slap tonguing	/slæp ˈtʌŋɪŋ/	弹舌
slide	/slaɪd/	拉管
slide brace	/slaɪd breɪs/	拉杆支撑
slide lock	/slaɪd lɑːk/	拉管锁环
sloped	/sloʊpt/	倾斜的
slur	/slɜːr/	连奏；圆滑音
snare	/sner/	响弦
snare guard	/sner gɑːrd/	响弦固定器
snare head/bottom head	/sner hed/, /ˈbɑːtəm hed/	共鸣面
soft palate	/sɔːft ˈpælət/	软腭
soft pedal (left pedal/ una corda pedal)	/sɔːft ˈpedl/	柔音踏板

solfeggio	/sɒlˈfedʒoʊ/	视唱练耳
sound post	/saʊnd poʊst/	音柱
soundboard	/ˈsaʊndbɔd/	音板
spatula key	/ˈspætʃələ kiː/	桌键
spiccato	/spɪˈkɑːtoʊ/	跳弓
spinto soprano	/ˈspɪntoʊ səˈprɑːnoʊ/	抒情兼戏剧女高音
Sprechgesang	/ˈʃprexɡəzæŋ/	朗诵唱；道白式演唱法
square wave	/skwer weɪv/	方波
staccato	/stəˈkɑːtəʊ/	断音；断奏；顿弓
staple	/ˈsteɪpl/	嘴套
Stevens grip	/ˈstivənz ɡrɪp/	斯蒂文斯握槌法
stool	/stuːl/	高脚凳
stradella bass	/strəˈdelə beɪs/	斯特德拉低音
strainer	/ˈstreɪnər/	上弦器；响弦器
strap	/stræp/	背带
strap ring	/stræp rɪŋ/	平衡杆；背带环
string	/strɪŋ/	琴弦
substantial	/səbˈstænʃl/	大量的
subtractive synthesis	/səbˈtræktɪv ˈsɪnθəsɪs/	减法合成
suspended cymbal	/səˈspendɪd ˈsɪmbəl/	吊镲
suspension	/səˈspenʃn/	控制气息
suspension ring	/səˈspenʃn rɪŋ/	悬挂环
sustain	/səˈsteɪn/	保持
sustain pedal (damper/ right/ resonance pedal)	/səˈsteɪn ˈpedl/	延音/右/共鸣踏板
switch	/ˈswɪtʃ/	键盘变音键
synchronize	/ˈsɪŋkrənaɪz/	同步
system module	/ˈsɪstəm ˈmɑːdʒuːl/	系统模块
tailpiece	/ˈteɪlpiːs/	系弦板
tambourine	/ˌtæmbəˈriːn/	铃鼓
temple block	/ˈtempl blɑːk/	木鱼
tempo	/ˈtempəʊ/	演奏速度
tenor joint/ wing joint	/ˈtenər dʒɔɪnt/, /wɪŋ dʒɔɪnt/	支管
tension casing/ lug	/ˈtenʃn ˈkeɪsɪŋ/, /lʌɡ/	鼓耳
tension rod	/ˈtenʃn rɑːd/	张力螺丝
the falsetto register/head voice	/ðiː fɔːlˈsetoʊ ˈredʒɪstər/, /hed vɔɪs/	头声 M2
the lower joint	/ðiː ˈləʊər dʒɔɪnt/	下节管

the modal register/chest voice	/ði: ˈməʊdl ˈredʒɪstər/, /tʃest vɔɪs/	胸声 M1
the upper joint	/ðə ˈʌpər dʒɔɪnt/	上节管
the vocal fry register	/ði: ˈvəʊkl fraɪ ˈredʒɪstər/	气泡音 M0
the whistle register	/ði: ˈwɪsl ˈredʒɪstər/	海豚音 M3
timbre module	/ˈtæmbər ˈmɑ:dʒu:l/	音色模块
timbre	/ˈtæmbər/	音色
tom tom	/tɑ:m tɑ:m/	嗵嗵鼓
tonguing	/ˈtʌŋɪŋ/	吐音
touch	/tʌtʃ/	触键
traditional grip	/trəˈdɪʃənl grɪp/	传统握槌法
transport	/ˈtrænspɔ:rt/	传送；输送
transposing	/trænˈspoʊzɪŋ/	移 / 转调的
treble clef	/ˈtrebl klef/	高音谱号
treble keyboard	/ˈtrebl ˈki:bɔ:rd/	键盘
tremolo	/ˈtreməloʊ/	震音；颤音（小提琴）；轮指
triangle	/ˈtraɪæŋgl/	三角铁
trill	/trɪl/	（指）颤音
trill key	/trɪl ki:/	颤音键
triple tonguing	/ˈtrɪpl ˈtʌŋɪŋ/	三吐
triplet	/ˈtrɪplət/	三连音
tube	/tu:b/	管体
tuning	/ˈtu:nɪŋ/	调弦
tuning indicator	/ˈtu:nɪŋ ˈɪndɪkeɪtər/	调音指示器
tuning machine	/ˈtu:nɪŋ məˈʃi:n/	弦轴
tuning pedal	/ˈtu:nɪŋ ˈpedl/	调音踏板
tuning peg/pegbox	/ˈtu:nɪŋ peg/, /peg bɑ:ks/	弦轴 / 弦轴箱
tuning rod	/ˈtu:nɪŋ rɑ:d/	调音杆
tuning slide	/ˈtu:nɪŋ slaɪd/	调音管
U-bend	/(j)u: bend/	U 形弯管
upper keyboard	/ˈʌpər ˈki:bɔ:rd/	上键盘
upstroke	/ˈʌpˌstroʊk/	挑
valve	/vælv/	活塞
valve cap	/vælv kæp/	活塞帽
valve slide	/vælv slaɪd/	活塞管
valve	/vælv/	气钮
velocity	/vəˈlɑ:səti/	力度
vibraphone	/ˈvaɪbrəfoʊn/	颤音琴

vibrato	/vɪˈbrɑːtəʊ/	（唇）颤音；揉弦（小提琴）
viola	/viˈəʊlə/	中提琴
violist	/viˈəʊlɪst/	中提琴手
vivace	/vɪˈvɑːtʃeɪ/	较活泼的速度；快板
vocal cord (vocal fold)	/ˈvəʊkl kɔːrd/	声带
vocal fold mucosa	/ˈvəʊkl fəʊld mjuˈkoʊsə/	声襞粘膜
vocal tract	/ˈvəʊkl trækt/	声道
vocalize	/ˈvəʊkəlaɪz/	发声；练声
voice teacher (or vocal coach)	/vɔɪs ˈtiːtʃər/	声乐导师
voicing	/ˈvɔɪsɪŋ/	音色
water key	/ˈwɔːtər kiː/	放水阀；水门
waveform	/ˈweɪvfɔːrm/	波形
wavetable synthesis	/ˈweɪvteɪbl ˈsɪnθəsɪs/	波表合成
xylophone	/ˈzaɪləfoʊn/	木琴
yodeling	/ˈjoʊdlɪŋ/	约德尔唱法；用真假嗓音交替歌唱